The Puzzling World of Polyhedral Dissections

RECREATIONS IN MATHEMATICS
Series Editor *David Singmaster*

1. Hugh ApSimon
   Mathematical byways in Ayling, Beeling, and Ceiling
2. John D. Beasley
   The Ins and Outs of Peg Solitaire
3. Ernö Rubik, Tamás Varga, Gerzon Kéri, György Marx, and Tamás Vekerdy
   Rubik's Cubic Compendium
4. L. E. Hordern
   Sliding Piece Puzzles
5. John D. Beasley
   The Mathematics of Games
6. Stewart T. Coffin
   The Puzzling World of Polyhedral Dissections

'The most beautiful thing we can experience is the mysterious. It is the source of all true art and science.'
Albert Einstein, *What I Believe* (1930)

The Puzzling World of Polyhedral Dissections

Stewart T. Coffin

Oxford New York
OXFORD UNIVERSITY PRESS
1990

Oxford University Press, Walton Street, Oxford OX2 6DP
Oxford New York Toronto
Delhi Bombay Calcutta Madras Karachi
Petaling Jaya Singapore Hong Kong Tokyo
Nairobi Dar es Salaam Cape Town
Melbourne Auckland
and associated companies in
Berlin Ibadan

Oxford is a trade mark of Oxford University Press

Published in the United States
by Oxford University Press, New York

British Library Cataloguing in Publication Data
Coffin, Stewart T.
The puzzling world of polyhedral dissections.
1. Polyhedra
I. Title II. Series
516'.15
ISBN 0–19–853207–5

Library of Congress Cataloging in Publication Data
Coffin, Stewart T.
The puzzling world of polyhedral dissections/Stewart T. Coffin.
p. cm.—(Recreations in mathematics; 6)
Bibliography: p.
Includes index.
1. Mathematical recreations. 2. Puzzles. I. Title. II. Series: Recreations in mathematics; no. 6
QA95.C59 1990 793.7'4—dc20 89–8683
ISBN 0–19–853207–5

Typeset by Cotswold Typesetting Ltd, Cheltenham
Printed in Great Britain by Courier International Ltd, Tiptree, Essex

Preface

This book had its conception in 1974, when my small cottage industry of designing geometrical puzzles and handcrafting them in wood was then in its fourth year. It began as a newsletter of very limited circulation having to do with mechanical puzzles in general, especially those that could be made in the classroom or workshop. It was intended for persons who enjoy designing puzzles, making them, collecting them, or just solving them. Its purpose was as much to gather information as to disseminate it. In 1978, the various issues were assembled into a booklet. Later, more chapters were added, and it was published as a book of sorts called *Puzzle Craft.* It was put together and bound right here in one corner of my workshop. A revised and improved edition was published in 1985.

In this Oxford University Press edition, the material has again been extensively revised and considerably expanded. The scope has been narrowed to focus on geometrical dissection puzzles, especially those that lend themselves to construction in the workshop or in the industrious factory of the imagination.

In addition to the numerous individuals whose valued contributions to this book are acknowledged at the appropriate places within the text, the following persons have contributed information or helped in some other way: Saul Bobroff, Leonard Gordon, Edward Hordern, Stanley Isaacs, Joseph Lemire, John Loeser, William Perkins, Thomas Rodgers, Louis Rosenblum, Norton Starr, and Roland Zito-wolf.

Kathy Jones translated my hopeless grammar and spelling into what I am assured is proper English and made many valuable suggestions, as did also Allan Boardman. Jerry Slocum checked the manuscript for historical accuracy. Special thanks also to David Singmaster and the staff of Oxford University Press.

Assisting in workshop and office were my wife Jane and, in bygone days, the three elves—Abbie, Tammis, and Margaret.

Stewart T. Coffin
Lincoln, Massachusetts

Contents

Introduction

Nearly everyone must have had at least a few amusements among his or her childhood treasures based on the simple principle of taking things apart and fitting them back together again. Indeed, many infants show a natural inclination to do this almost from birth. Constructing things out of wooden sticks or blocks of stone must surely be one of the most primitive and deeply rooted instincts of mankind. How many budding engineers do you suppose have been boosted gently along toward their careers by the ever-lasting fascination of a mechanical construction set? I know I certainly was. Even after life starts to become more complicated and most childhood amusements have long since been left by the wayside, the irrepressible urge to join things together never dies out. (Quite the contrary!)

Construction pastimes in the form of geometrical assembly puzzles have a universal appeal that transcends all cultural boundaries and practically all age levels. Young children catch on to them most quickly. One of the puzzle designs included in this book was the inspiration of an eight-year-old, and children younger

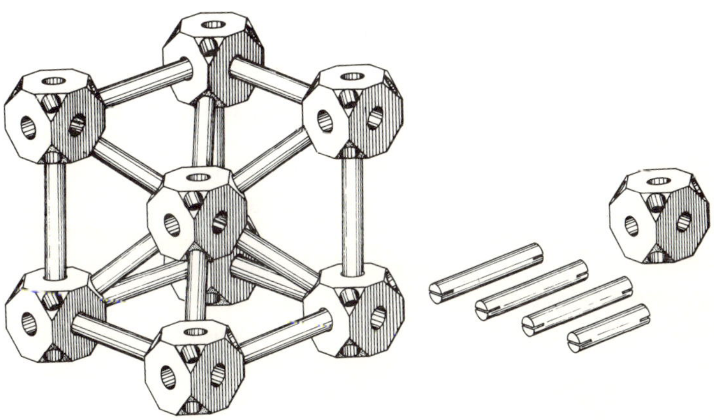

Fig. 1 *'He who wonders discovers that this is in itself a wonder.'* M. C. Escher

than that have solved many of them. So much, then, for the presumptuous practice of rating the difficulty of puzzles according to age level, with adults of course always placing *themselves* at the top! Likewise, almost anyone from elementary school student to retiree having access to basic workshop facilities should be able to fabricate many of the puzzles to be described on the following pages.

On the other hand, this book is intended to be more than simply a collection of puzzle designs, plans and instructions. This is a puzzle designer's guidebook. Some of the most rewarding recreations are neither in simply solving puzzles nor in making them, but rather in discovering new ideas and crafting them into a form that others may enjoy too. Equally satisfying is to discover surprises long overlooked in traditional puzzles. It is amazing how many of these lie scattered about just beneath the surface waiting to be uncovered. Keep in mind that the systematic investigation of many types of problems covered in this book has taken place only within the last decade or two. Throughout these pages, unsolved problems are mentioned, or at least implied, that should keep mathematicians and analysts, tinkers and inventors occupied for a long time to come. A few gems have even been purposely reburied so that the reader may have the joy of unearthing them again. But watch out for traps!

Life in general is a puzzle, is it not? Examples abound—trying to fathom the mysterious rules of English grammar and wondering if the spelling of some words was someone's idea of a joke! The engineer who dreamed up the assembly procedure for my car's transmission passed up a promising career as a puzzle inventor. Anyone who writes poetry or composes music knows the satisfaction that comes when all of the parts finally fit together properly, or the frustrations when they decline to. Almost any undertaking may become turned into a puzzle, intentionally or otherwise.

This book is devoted to a broad and vaguely defined classification of geometrical recreations that might be described as *burrs* and *polyhedral dissections.* Polyhedra are by definition any solids bounded by plane surfaces. One often associates the term *polyhedra* with the isometrically symmetrical Platonic solids and their relatives. It is used here in a broader sense to include practically any solid or assemblage of parts having some sort of symmetry, including burrs. In puzzle nomenclature, burrs are assemblies of interlocking notched sticks. They are traditionally square sticks, but all sorts will be considered here.

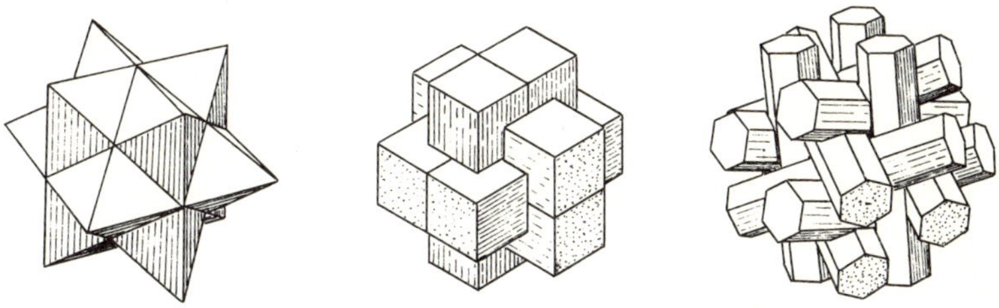

Fig. 2 Burrs

For convenience, the term *puzzle* is used throughout this book to include just about any sort of geometrical recreation having pieces (actual or imagined) that come apart and fit back together again. Probably many readers associate the word *puzzle* with some task that is purposely confusing or difficult. That notion may be rather misleading when applied to many of the recreations described in this book. I much prefer to regard them as being fascinating and intriguing. Discovering the myriad amazing ways in which geometrical solids fit together in space is in itself a marvellous revelation. If they also have the potential for challenging puzzle problems, so much the better. But let us not make the common mistake of assuming that the more satisfactory puzzle is one that is fiendishly difficult or complicated—a tendency more often than not counterproductive in any creative endeavour.

A proper treatise on geometrical puzzles should probably begin with a historical overview. Here we have a problem. If you search long enough, you can usually find at least a brief written history on just about any possible subject, but apparently not so for geometrical puzzles. Likewise, a search through the major anthropological museums of the world turns up practically nothing. The conclusion to be drawn from this is that most puzzle designs must not be very old—almost none over 200 years. A popular marketing ploy of puzzle manufacturers is to invent stories of their ancient origins. One favourite theme is that they came down to us from the Orient. Some authors have called the six-piece burr the *Chinese Cross* puzzle. Conversely, perhaps puzzles sold in Japan are touted as products of Yankee ingenuity, and if so, probably they are closer to the truth.

Patent files are one of our most important historical resources on puzzles. There are presently about 1000 patents of bona fide puzzles filed in the US Patent Office and about the same number in the British Patent Office. The oldest US patent is dated 1863. If the filing of patents is any accurate indication, then many of the classic designs familiar to us today, including various burrs and dissected blocks, date from the late 1800s. Starting around 1920, there is a decline in puzzle interest and patent activity (which by the way just happens to coincide with the phenomenal rise in popularity of the automobile). Puzzle interest picks up again after World War II and has been going strong ever since.

Many games and pastimes are known to be quite ancient, so why not three-dimensional puzzles too? We can only speculate, but here is one thought: of all three-dimensional puzzles, the so-called burr or notched square stick types are certainly the most familiar, the easiest to make, and probably the earliest to have become popular. To be entirely satisfactory, such puzzles should be made to close tolerances, and the only practical way to achieve this is with specialized power woodworking machinery and suitable jigs. Power woodworking tools did not come into common use until the mid-19th century. Note that most ancient games and pastimes use pebbles, beans, scratch marks on the ground, and other such things readily at hand.

To say that most geometrical puzzles are less than 200 years old requires qualification. They are all based on mathematical principles known ages ago, which in turn have roots going even further back, finally fading away into the unknown of the past. To give credit where it is most due, the fascinating world of geometrical dissections, and indeed of mathematical recreations in general, is utterly and profoundly Greek in origin. Behind every geometrical model illustrated in this book, the shadow of the Acropolis looms dimly in the background; and within every tortuous puzzle solution lurks the ghost of the fabulous labyrinth of King Minos, brooding over its next victim!

The term *mathematical recreation* is in itself rather a misnomer, for every geometrical puzzle worthy of consideration has non-mathematical aspects that are just as important if not more so. Most of the puzzle ideas described in this book were conceived by someone who was not a mathematician by either training or

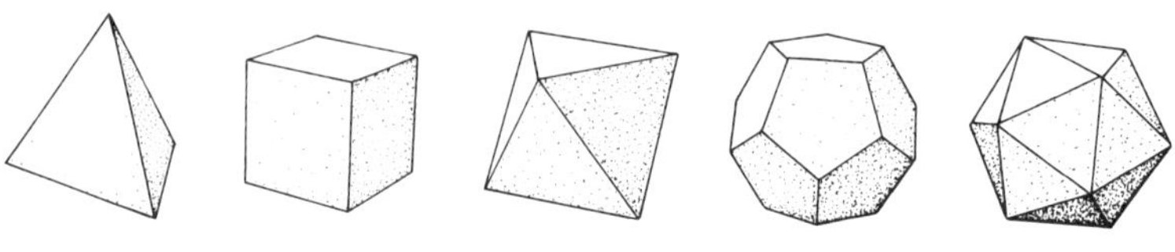

Fig. 3 The Platonic solids

profession but rather more of an inventor and craftsman, with perhaps a whimsical or artistic bent. Conversely, many creative endeavours that we certainly do not regard as geometrical puzzles involve essentially the fitting together of discrete parts artistically into a logical and harmonious interlocking whole. The aspiring puzzle inventor seeking guidance and inspiration in the art of invention may discover more of it in one Bach cantata than in all the world's mathematics textbooks.

Except for this edition's predecessor, *Puzzle Craft*, there have been virtually no books ever written specifically on geometrical puzzles. Many books on mathematical recreations have touched on the subject. There have been several compendia of mechanical puzzles in general that have included some burrs and geometrical dissections. Likewise a few woodworking books have included a chapter or two on puzzles. The closely related subjects such as polyhedra, symmetry, combinatorial theory, and design science all have extensive literature. Perhaps it is inherent in the very nature of dissection puzzles that even their literature is thus so scattered in bits and pieces. Trying to fit all of them together for the first time is quite a puzzle in itself!

Until recently, puzzles were regarded as little more than children's toys, and certainly not as a subject worthy of university-level study or museum exhibits. Before World War II, many wooden puzzles were mass-produced in the Orient, using the same few simple designs year after year. Typical were those found in the illustrious Johnson Smith & Co. mail order catalogue of the 1930s (Fig. 4), priced at 10 cents or 15 cents postpaid! Then cheap plastic

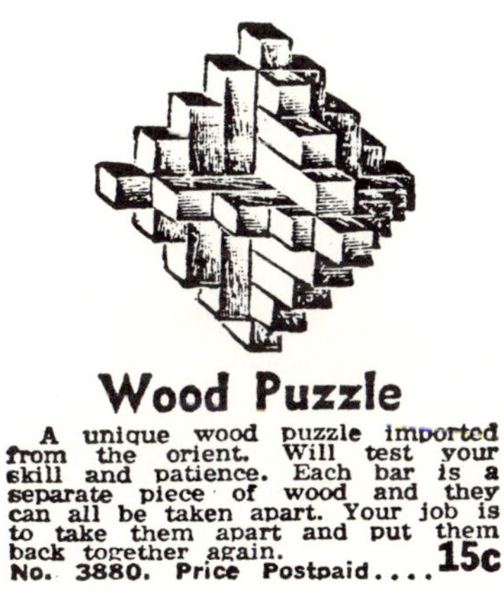

Fig. 4

versions in injection-moulded styrene started flooding the market, perpetuating the image of puzzles as expendable toys and trinkets. But all that is changing. There is a growing interest in geometrical recreations at all levels, from educational materials for pre-schoolers to university courses and seminars, arts and crafts exhibits, articles in scientific journals, and yes, even a few good books!

One reason that geometrical dissections have so much potential for recreation is the wide range of skills and talents that may be brought into play, from the theoretical to the practical and from the mathematical to the artistic. At the practical level, a complex interlocking puzzle well crafted in fine wood can be a challenging and rewarding project for the skilled woodworker. On another level, some persons are more intrigued by the geometrical shapes themselves, and a sort of Greek renaissance subculture has sprung up in the field of architecture and decorative design having to do with the adoration of polyhedra. On yet another level, there is what I call, for lack of a better term, the psycho-aesthetics of puzzle design. This gets into the puzzling question of what it is that makes certain puzzles appeal to certain persons but not others. So far as I know, almost nothing has previously been written on this pregnant subject.

Most of the designs described in this book are for puzzles that can, in theory at least, be made in wood. Directions and helpful hints for doing so are given. Some are much easier to make than others. You can start with the easy ones and gradually work upward, depending upon your woodworking skills and workshop facilities. But what about the reader with no such inclination or no workshop? Do not despair. Many of the designs have been or are being produced commercially, and probably many more will be in the future. Furthermore, the reader with good spatial perception ought to be able to solve many of them visually or on paper, without the need for physical models.

We might carry this notion a step further and suggest that the essence of an intriguing geometrical puzzle is really the idea behind it. The physical model of the puzzle then becomes more of a tool to aid the thinking process and help convey the idea. Crude models may suffice for this purpose. As you become more adept with these skills, you may find that the actual models assume less importance than the principles involved. Some designers and solvers of geometrical puzzles work almost entirely in the abstract, using pencil and paper or a computer, plus the amazing imaginative powers of the human mind. Consider all the advantages: the parts always fit perfectly and, unlike their wooden counterparts, never swell or shrink, crack or break. And for the apartment dweller with limited space, just think how many designs can be created and stored inside the recesses of one's head, using spaces that might otherwise have remained vacant!

Most of those who invent puzzles like to be given credit when their ideas are published, and some even hope to profit from them. Mention is made of the originators or patent grantees for a few of the puzzles described in this book when known, especially for some of the older classics. Well over half of all the designs included in this book were conceived and published only within the past 15 years. Although the origins of most of them are known to the author, credit is purposely omitted for these reasons: some of the ideas are so obvious that they probably have been discovered independently by more than one person. Others may be just minor variations of someone else's ideas. For example, one of the puzzles described in this book is the author's variation on a design picked up recently from another puzzle craftsman in Florida, who reports getting the idea from someone in California, who in turn reports getting it from a puzzle company in Europe. But the idea is said to have originated in Japan, although it too is but a variation on a familiar theme. An analysis of its solutions came to me from yet another source, and he reports learning that someone else had done it independently. Trying to unravel something like that would perplex even a patent attorney. So, some of the puzzles in this book are in the public domain, some are patented, some are copyrighted, and some are none of these. But the author cannot say in all cases which are which, so will avoid misunderstandings by not trying to. Anyone planning to manufacture or publish any of them should undertake the research necessary to make certain that no one's rights or sense or pride are being overlooked.

Chapter 1

Two-dimensional dissections

Most of the designs described in this book can be thought of as dissections of some sort. By way of introduction, we will first consider some simple two-dimensional geometrical dissections, which in their physical embodiment become assembly puzzles.

Jigsaw puzzles

To dissect means literally to cut into pieces. Just about any chunk of material cut into pieces becomes a sort of dissection puzzle. If sawn freely perpendicular to the surface of a sheet of plywood (or die-cut of cardboard), the result is the familiar jigsaw puzzle. Most jigsaw puzzles are not designed to exercise or perplex the mind, at least in the sense that other types of puzzles do, and it is perhaps stretching the definition a bit to even call them puzzles. The definition given in the dictionary for the noun *puzzle* seems to have been purposely broadened so as to include what are really pastimes of pattern recognition, memory, and patience. The definition given for the verb *to puzzle* contains no such connotation.

Jigsaw puzzles have been around for over 200 years, longer than nearly any other type of puzzle. Although their relationship to burrs and polyhedral dissections may appear to be remote, they are probably an important historical root. The ancestry of inventions in general must be an incredibly complex web of ideas branching backward in time into just about every nook and cranny of human culture. Puzzles are certainly no exception, and jigsaw puzzles, by their sheer numbers and long history, must play at least a minor role in the evolution of many present-day geometrical puzzles and recreations. How many of us played with jigsaw puzzles at one time and then began to ponder, perhaps subconsciously, variations along logical and mathematical lines?

Various schemes have been employed to make jigsaw puzzles more clever, such as by sawing on two different faces of a rectangular block or along multiple axes of a sphere. Some of these are quite entertaining, but still they are essentially non-geometrical in principle.

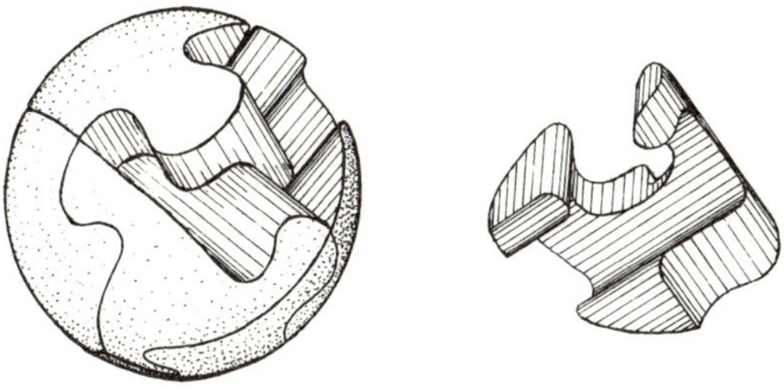

Fig. 5

Tangram

If, instead of cutting freely, the dissection is done according to some simple geometrical plan, an entirely different type of puzzle results. Many fewer pieces are required to create interesting puzzle problems. Three characteristics of such puzzles are that they nearly always use straight line cuts, they usually assemble into many different puzzle shapes, and the problem shapes often have more than one solution.

Of the types of puzzles covered in this book, the oldest known is the popular seven-piece dissection of the square known as Tangram. It was at one time thought to be thousands of years old, but is now known to have originated in China sometime before 1780. (A quite similar Japanese seven-piece square dissection has been dated back to 1742.) Tangram became popular throughout Europe and America in the 19th century and continues to be so to this day. It is made and sold in many different materials. Thousands of problem shapes have been published for it over the years, and it is mentioned in many books. For more background information on Tangram and many similar puzzles, the reader is referred to *Puzzles Old and New* by Botermans and Slocum. Here we will discuss some of the curious mathematical aspects of the puzzle not generally mentioned in the literature. The dissection is shown in Fig. 6.

In designing dissection puzzles of this type, the idea is to divide the whole according to some simple geometrical plan so that the pieces will fit together many different ways. The way this is accomplished in Tangram is shown in Fig. 7. A diagonal square grid is superimposed onto the square whole such that the diagonal of the square measures four units and the area is eight square units. The only lines of dissection allowed are those that follow the grid or diagonals of the grid. To put it another way, the basic structural unit is an isosceles right-angled triangle made by bisecting a grid

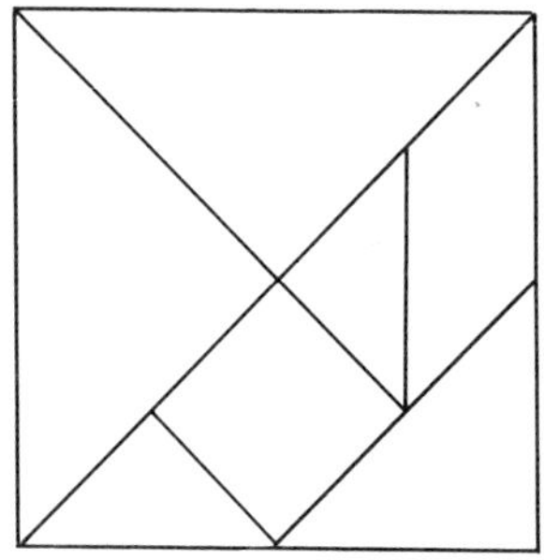

Fig. 6

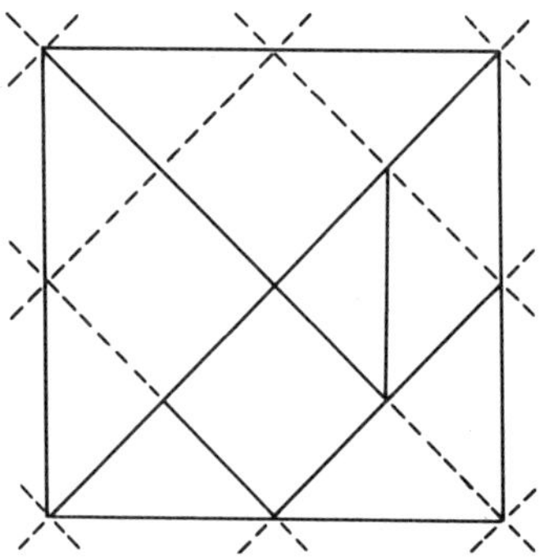

Fig. 7

square, and all larger puzzle pieces are composed of these unit triangles joined together different ways. In Tangram, there are two of the unit triangles alone, three pieces made up of two unit triangles joined all possible ways, and two large triangles made up of four unit triangles, for a total of 16 unit triangles. The relative lengths of all edges are thus powers of $\sqrt{2}$.

The first Tangram problem is to scatter the pieces and then reassemble the square. Note that it has only one solution, usually a mark of good design. (Rotations and reflections are not counted as separate solutions.) For the countless other problem shapes, you can try to solve the published ones found in many books and magazines, or you can invent your own.

The easiest way to discover Tangram patterns is just by playing around with the pieces. Start by trying to make the simplest and most obvious geometrical shapes—triangle, rectangle, trapezoid, parallelogram, and so on, always using all of the pieces. An alternate method is to draw some simple shape on graph paper following the rules already given and having an area of eight squares, and then try to solve it. Which of the examples shown in Fig. 8 are possible to construct?

Fig. 8

Published Tangram patterns range all the way from the geometrical shapes shown above to the other extreme of animated figures created by arranging the pieces artistically. This range is represented by the row of figures shown in Fig. 9, reading left to right. Only those solutions that conform to a regular grid can be considered true geometrical constructions. Careful inspection will show those to be the three on the left. The others may be very artistic and imaginative, but they are not within the province of this book.

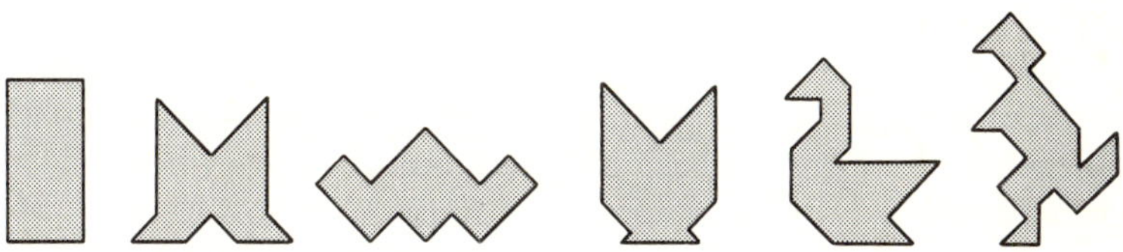

Fig. 9

The theme of discrete rather than random or incommensurable ratios of dimensions is one that plays continuously in the background throughout this book. In the case of Tangram-like dissection puzzles, it is easy to see that they cannot be made to work properly any other way. Beyond that, though, there must be something inherently appealing to our aesthetic sensibilities in simple, discrete ratios. They are, after all, the foundation of all music, although probably no one understands exactly why.

Fig. 10 shows 13 convex Tangram pattern problems. A *convex* pattern is one that can be cut out with a paper cutter straight away, i.e. with no holes or inside corners. They are all possible to construct. Are any others possible?

For a slight change of pace from the usual Tangram problem, consider the following puzzler, which by the way is based more or less on an actual happenstance: Karl Essley made two Tangram sets

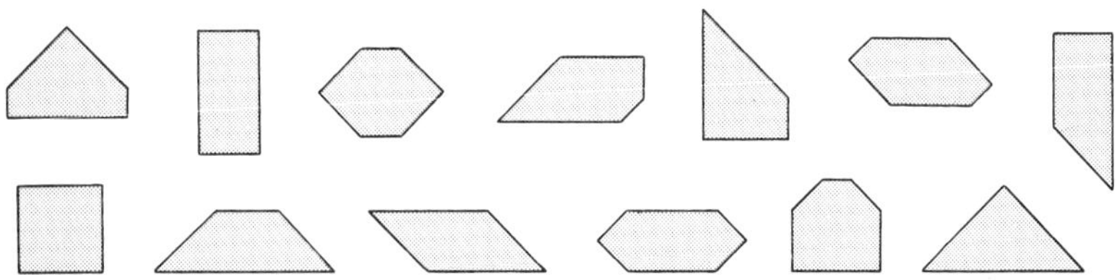

Fig. 10

as gifts—one to be sent to his sister and the other to his brother. The instructions were simply to assemble all the pieces into a square. Karl's sister brought hers back and declared (correctly) that the solution was impossible. Examining her set, they discovered that Karl had made a mistake in packing and had accidentally put two pieces into the wrong box, so one person got a set of five pieces and the other got nine. Embarassed, Karl suggested that they phone their brother and explain the mistake. But his sister reflected for a moment and then said, 'No, that won't be necessary—he *can* make a square with *his* set.' Can you tell who got the two extra pieces, and what shape or shapes they were? (Answer later in book.) Be careful—this puzzler contains a nasty trap!

In a similar vein to the above puzzler, note the pairs of figures shown in Fig. 11. In each pair, one figure appears to be complete and the other appears to have a piece missing; yet they both use all seven pieces, as all Tangram figures must. Can you discover the common characteristic that all such confusing pairs have? What other such pairs can you discover?

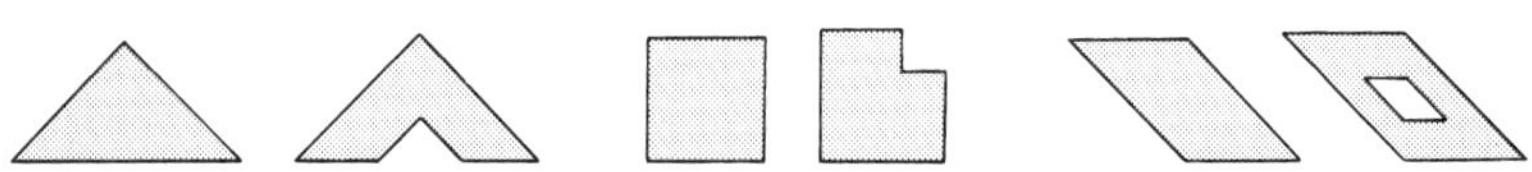

Fig. 11

In order to be entirely satisfactory, especially considering the examples just given, even simple puzzles such as this one should be accurately made of stable materials. If sawn directly out of a square of plywood, there will be noticeable errors introduced by the saw kerf. A more accurate way is to lay it out on cardboard, cut the cardboard with scissors, and then use the cardboard pieces as patterns.

Throughout this book, unscaled drawings are given for puzzle constructions. There are always a few readers who will report being unable to use such drawings, having been indoctrinated in school with the notion that nothing can be made out of wood without standard workshop blueprints with dimensions. Dimensions are omitted for the following reasons:

1. They are unnecessary. It should be obvious, for example, that in Tangram all of the angles are 45 or 90 degrees.

2. They are not as accurate as geometrical constructions. If the overall Tangram square is integral, all of the diagonal measurements are irrational and can be expressed in sixteenths of an inch or whatever only by rounding off.

3. Adding practical dimensions would only tend to obscure the elegantly discrete mathematical essence of the problem with unessential detail.

4. You may scale the puzzle to any size you wish.

Other Tangram-like puzzles

The great popularity of Tangram has spawned many imitations. Most notable of these were the famous Anchor Stone puzzles produced by Richter and Company of Germany, starting in the late 1800s and on into the early 1900s. In *Puzzles Old and New*, Botermans and Slocum show 36 different designs, and some of these are worth examining. Six of them, including Tangram, are squares dissected according to the usual square grid with diagonals. Three of these, however, are on a grid with a finer scale than Tangram, i.e. containing more grid squares and unit triangles. The diagrams in Fig. 12 should make this clear. The number below each

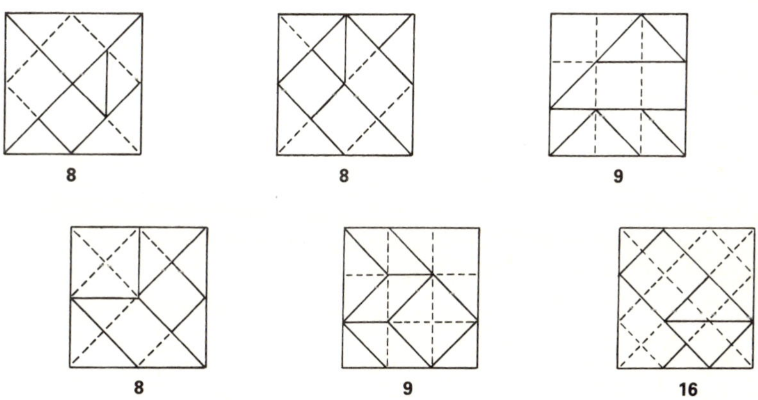

Fig. 12

one indicates the number of grid squares enclosed for the coarsest grid that will conform.

For a given number of pieces, dissections with coarser grids are likely to have more mutually compatible edges—thus the three on the left in Fig. 12 are the better designs in this respect. A dissection that accomplishes its purpose with the fewest pieces is usually to be preferred—thus the two on the left in Fig. 12 emerge as the better designs. The final test is to see which of these two sets constructs more interesting puzzle figures, and this task is left to the reader. The one on the far left is of course Tangram, and the other one was sold under the name *Pythagoras.*

Incidentally, note that the next smaller possible grid would contain only four squares and eight unit triangles. Are these too few to make an interesting puzzle? The most obvious such set would be Tangram with the two large triangles omitted. This simple little set of five pieces probably contains a treasure-trove of undiscovered recreations. For example, how many convex patterns will it form?

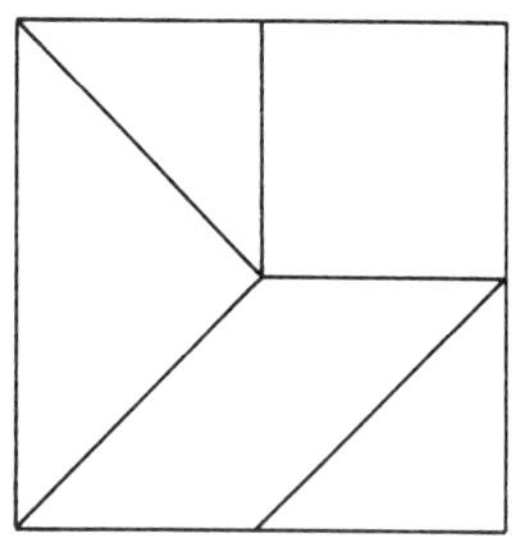

Fig. 13

A square can be dissected into numbers of equal isosceles right-angled triangles given by the following series: 2, 4, 8, 16, . . . What is the next number in this series? (This question is reminiscent of 'IQ' tests school children used to be given, and probably still are. Example: given the series 4, 6, 8, . . . , what is the next number? A precocious student interested in prime numbers might answer *9,* while one intrigued by the Platonic solids might say *12.* But of course, by this time the students are supposed to know that the way the system works is to always give the answer that the *teacher wants,* no matter how uninspired!)

Next in the Richter series, we find eight puzzles similar to those in Fig. 12 except rectangular rather than square. These are shown in Fig. 14 without further comment, except to point out that puzzles

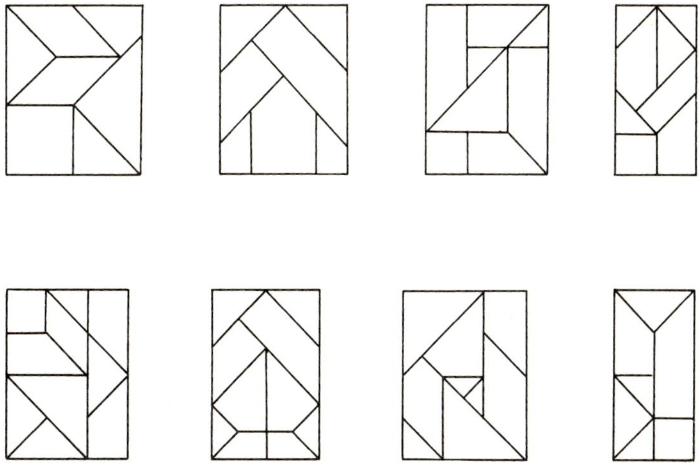

Fig. 14

with mostly dissimilar pieces are generally more interesting than those with many duplicates or triplicates.

All of the Richter puzzles shown so far have used only 45- and 90-degree angles. Eight of the Richter puzzles are polygonal shapes dissected into pieces with 30-60-90-degree angles. These are shown in Fig. 15, arranged by increasing numbers of pieces.

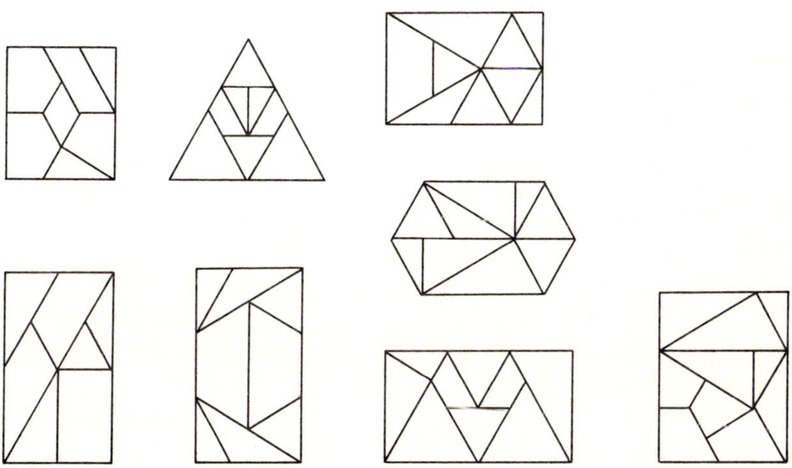

7 8 9 10

Fig. 15

Most of the other Richter puzzles have curved outlines or other complications. For example, the two shown in Fig. 16 have more complicated angles. In dissection puzzles of this type, if all of the angles and linear dimensions are not immediately obvious by inspection, then the design is probably not very well conceived.

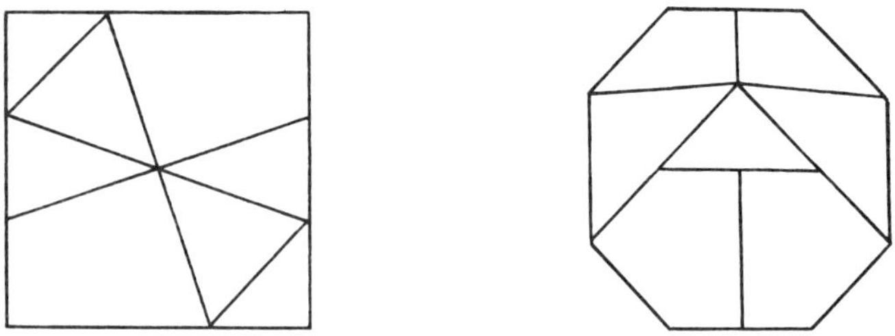

Fig. 16

To digress slightly, a most curious dissection is the one shown in Fig. 17 on the left. This construction within a square appears in *Curiosités Géométriques*, by E. Fourrey, published in Paris in 1907. It is said to have been discovered in a 10th-century manuscript and is supposed to have been the work of Archimedes. At least three slightly different versions of it have appeared in modern puzzle books, all supposing it to be a geometrical dissection puzzle and calling it the 'Loculus of Archimedes'. One learns to be sceptical about such things, especially when they do not appear to make much sense and the original documents are reported lost. The mystery of its origin and its actual purpose is a challenging problem for recreational maths historians.

It has been pointed out by some authors that the areas in the Loculus are cleverly devised to be in the ratios of whole numbers,

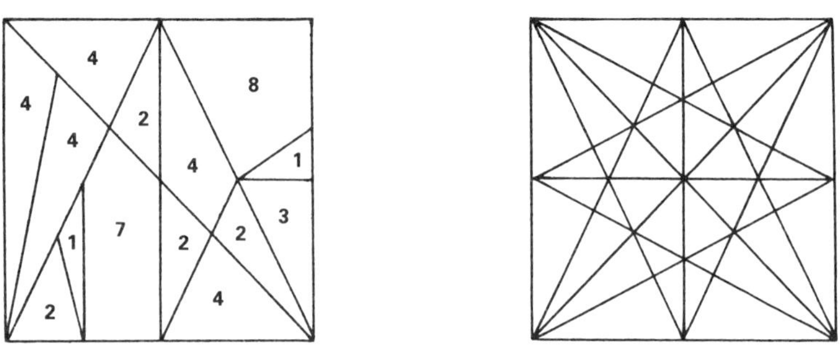

Fig. 17

as indicated. But there is nothing unusual about that. It is easily proven, if not immediately obvious, that all polygons formed by connecting points on a regular square grid must have areas in the ratios of whole numbers. Less obvious but also provable is that polygons formed by intersections of such lines must also have this property, as in the example shown in Fig. 17 on the right. Exercise for the reader: compute the relative areas in this figure.

Note that none of the Richter puzzles has fewer than seven pieces, and several have more. One always tries to minimize the number of pieces without sacrificing other design objectives. Satisfactory dissection puzzles of this type with fewer than seven pieces are not as common, but possible. Consider the experience of another puzzle acquaintance of mine, Bill Trong. Bill made for himself a Tangram set from published plans, but he carelessly failed to make one cut, so he ended up with two of the pieces joined together and thus a set of six pieces. Surprisingly, he found he could construct all 13 of the convex patterns (Fig. 10) with this set. Which two pieces were joined together? Judge for yourself if this six-piece version is an improvement over the original Tangram.

On page 11, the reader was asked if other convex Tangram solutions could be found. According to an article in *American Mathematical Monthly*, vol. 49, in 1942 Fu Traing Wang and Chuan-Chih Hsiung of the National University of Chekiang proved that no more than 13 different convex Tangrams can be formed. Their proof involved showing that there are only 20 possible ways of assembling the 16 unit triangles convexly, of which 13 were found to have Tangram solutions. An excellent discussion of this is given in *Tangram*, by Joost Elffers.

The point to be made here, before leaving the subject of

Fig. 18

Tangram, is that the simplest and most familiar puzzles often contain surprising recreational potential, much of which may have been overlooked. Some of the practical innovations may be quite clever too. Figure 18 shows an example of what one skilled and inspired woodcraftsman—Allan Boardman—has done with Tangram. The seven pieces fit with watchmaker's precision two layers deep into the tiny square box, complete with sliding cover, all beautifully crafted of pearwood.

A five-piece square dissection

Fig. 19 shows Sam Loyd's well known square-dissection puzzle. It is made by locating the midpoints of all four sides of the square, drawing the appropriate lines, and dissecting. The five pieces construct all of the puzzle patterns shown. Again, note the interesting paradox of the two on the right—one being a solid rectangle and the other a rectangle with a corner missing, yet both use all five pieces!

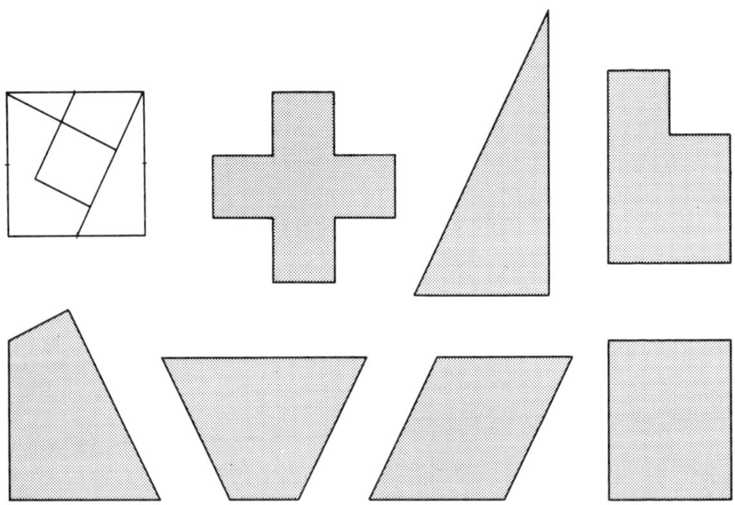

Fig. 19

When one of Loyd's pieces is divided in two, the number of possible interesting puzzle patterns is approximately doubled. Some of these new patterns are shown in Fig. 20. The first problem for the reader is to discover the additional cut. It should be obvious which piece to divide, but which way?

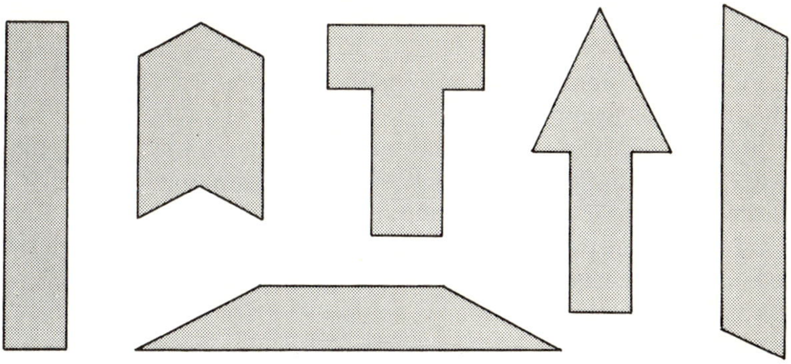

Fig. 20

The reader is now encouraged to experiment with new and original dissection puzzles. Start with a simple shape such as a square or rectangle and dissect it according to some simple geometrical plan, the idea of which is to make pieces that fit together many different ways. Six or seven pieces is a good number. Try to avoid having many pieces alike. Then create your own catalogue of pattern problems.

Geometrical dissections

To mathematicians, the term *geometrical dissection* has a slightly different meaning from the one we have been using here. It usually refers to two different polygons being formed from the same set of pieces. This is essentially an analytical problem, and a minor branch of mathematics is devoted to it. It has been proven that any polygon can be dissected to form any other polygon of the same area. Most attention has been given to the *regular* polygons. Choose any two regular polygons, and cut one of them into as many pieces as you wish to form the other. It may sound easy until you actually try it!

The classic problem in geometrical dissections is to find the *minimum* number of pieces required to perform a dissection between various pairs of common polygons. An excellent book on the subject is *Recreational Problems in Geometrical Dissections and How to Solve Them*, by Harry Lindgren.

Famous puzzle inventor Henry Dudeney was a pioneer in geometrical dissections. His classic four-piece dissection between the square and equilateral triangle, first published in 1902, is shown in Fig. 21. This must be the simplest of all possible dissections between two regular polygons. Yet if the reader will try to construct the dissection, even after glancing at the drawing, it will immediately become obvious that the methods described earlier in this chapter *do not work!*

Start by constructing a square and equilateral triangle of equal area. Thus, if the square is 1×1, the sides of the triangle are $2/\sqrt[4]{3}$. Next, note that all points marked (*) are midpoints of sides. Therefore, triangle ABC is equilateral, and point B on the square is located by measuring $1/\sqrt[4]{3}$ from point A, after which the rest is obvious.

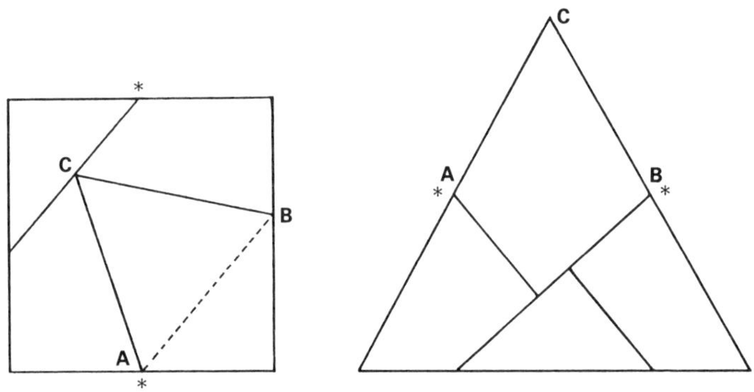

Fig. 21

In geometrical recreations of this sort, the essence of the puzzle is discovering the dissection. Given the dissections, their physical embodiment in the form of actual puzzle pieces has never enjoyed much popularity as practical manipulative puzzles. Perhaps it is because the two solutions are quickly memorized, and then there are no more problems. But there are exceptions. The Sam Loyd dissection puzzle described in the previous section was most likely developed by dissecting the square into the cross, after which the other interesting problem shapes were probably discovered. *Creative Puzzles of the World*, by van Delft and Botermans, contains an excellent chapter on geometrical dissections as practical puzzles. Further investigation might uncover a dissection by which *several* polygons could be constructed from a neat set of pieces. For example, what are the fewest pieces required to construct *three* different *regular* polygons? (Answer unknown, at least to the author.)

Checkerboards

Checkerboard puzzles consist of a dissected standard 8 × 8 checkerboard (draughtsboard). The object is not only to reassemble the pieces into an 8 × 8 square, but to do so with the proper checkering. A compendium of checkerboard dissections compiled by Jerry Slocum in 1983 lists 33 different versions, and it includes only those that have been manufactured, patented, or published. The numbers of pieces range from 8 to 15, with 12, 13, and 14 being the most common. The oldest is dated 1880. The commercial versions were usually made of die-cut cardboard printed on one side only, so the pieces may not be flipped. Some are printed on both sides, and the checkering may not be the same on both sides. Those made of light and dark wooden squares can of course be flipped. A typical 12-piece dissection taken from Slocum's *Compendium* is shown in Fig. 22. The pieces may not be flipped. It is known to have at least two solutions.

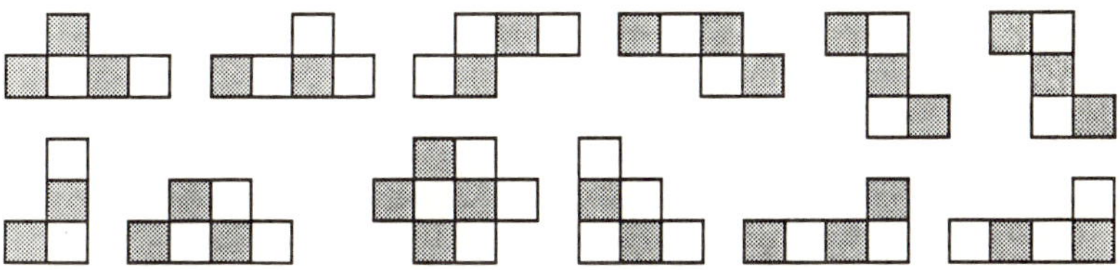

Fig. 22

Taken as a whole, checkerboard dissections tend not to be the most inspired of puzzle designs. All that can be said for most of them is that they differ slightly from each other. Any reader wishing to make a checkerboard dissection puzzle might just as well create an original design rather than copy someone else's. Here are some design suggestions:

1. As the number of pieces is increased, the difficulty increases, reaches a maximum, and then diminishes. For the checkerboard, maximum difficulty occurs around 11 or 12 pieces.

2. Difficulty of finding one solution varies inversely with the number of solutions possible. Designs with only one solution are considered especially clever (but how do you know?).

3. Pieces with compact shapes approximating square or rectangular, such as those containing a 2 × 2 square, lend themselves more easily to solutions and increase the number of

solutions. Contrarily, skinny, angular, complicated shapes do just the opposite, especially those that refuse to fit into corners.

4. To be avoided are pieces having rotational symmetry, and especially pieces identical to each other. (There will be more on this later. For a simple explanation here, imagine a checkerboard dissection in which this rule is grossly violated and see how exceedingly uninteresting it would be.)

It is interesting to note that the additional constraint imposed by the checkering may make the solution (or solutions) easier or harder, depending upon the circumstances. If only one mechanical solution exists to begin with, obviously the checkering makes it much easier to find. On the other hand, if hundreds of solutions exist, but only one with the correct checkering, then the addition of the checkering has turned it into a real puzzler!

Chapter 2

Two-dimensional combinatorial puzzles

A combinatorial problem (puzzle) is one in which various elements (pieces) can be combined (assembled) many different ways, only a few of which are the desired result (solution). The success or lack of it for any attempt at solution may not become apparent until most of the pieces are in place. For a geometrical puzzle, ideally all pieces are dissimilar and non-symmetrical, thus resulting in the maximum number of combinations for a given number of pieces. Maximum difficulty is achieved when only one correct combination exists. Since puzzles of this type can usually be made more difficult simply by increasing the number of pieces, the challenge facing the puzzle designer is to cleverly devise simple puzzles of this sort having few pieces while yet being intriguing and puzzling. In this chapter, we will introduce the subject by considering some simple two-dimensional combinatorial puzzles.

Regular polygons as building blocks

The basic building block of a geometrical combinatorial puzzle is typically a regular polygon, although other shapes or combinations of shapes are certainly possible. Whatever shape or shapes are used, the idea is to create a set of dissimilar puzzle pieces that fit together a great many different ways. Among regular polygons, the only ones that tile the plane are the triangle, square, and hexagon (Fig. 23).

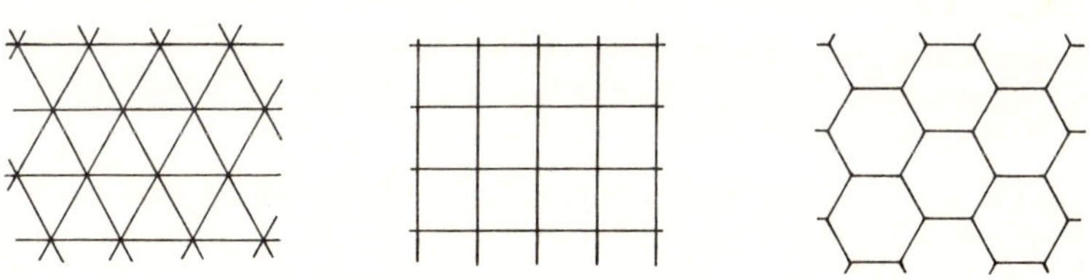

Fig. 23

Triangles as building blocks

Fig. 24 illustrates all the different ways of joining triangles through size-six. Note that mirror-images are not counted as separate pieces, since it is assumed these are real physical pieces that can be flipped over. These pieces are sometimes referred to as *polyiamonds.* The numbers of pieces are summarized in Table 1.

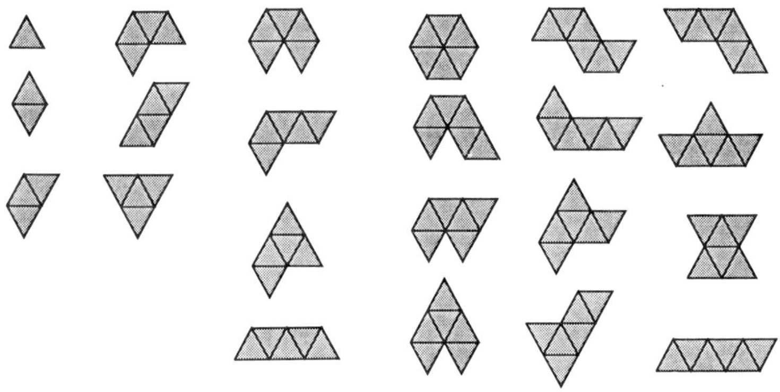

Fig. 24

Table 1

Size	Number of pieces	Total number of blocks
1	1	1
2	1	2
3	1	3
4	3	12
5	4	20
6	12	72

The next step is to consider what simple geometrical shapes these pieces might be assembled into. For the triangle, the most obvious patterns are triangular and hexagonal. These are shown in Fig. 25 in increasing size.

Comparing the numbers in Table 1 with the total numbers of blocks in different-sized sets, we note that none of them match. At this point, there are two different schools of thought. Those whose interest is primarily mathematical analysis like to work with *complete* sets of things, so they would probably either tinker with the definitions of the sets in an attempt to make the numbers match

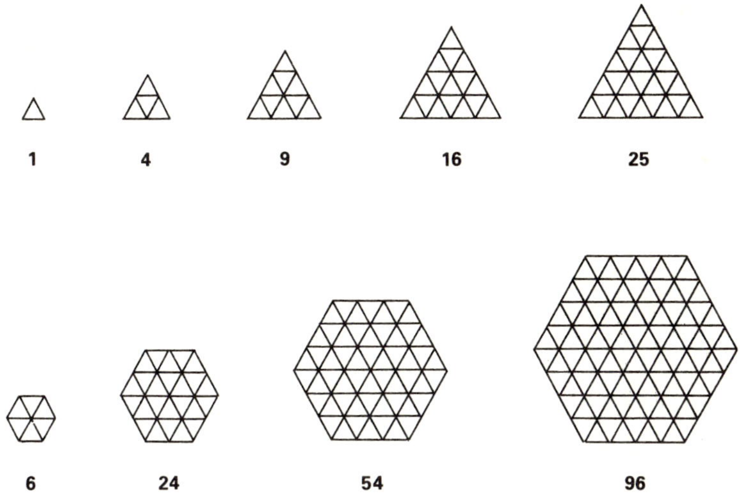

Fig. 25

or perhaps abandon this particular line of inquiry. From the practical point of view, on the other hand, there is no good reason why the pieces of a puzzle must comprise a complete mathematical set in order to be interesting. The tabulation of complete sets is useful in that it shows all the pieces that are available without duplication. Example: select a set of nine six-block pieces that assembles into a size-54 hexagon. Second problem: does such a set exist having a unique solution? (Answer unknown).

For those who do insist on working with 'complete' mathematical sets, note that of the 12 pieces of size-six, nine of them can be made by joining three two-block rhombuses together all possible ways, as shown in Fig. 26. Of course, this also has practical woodworking significance. Assemble these nine pieces into a 54-block hexagon.

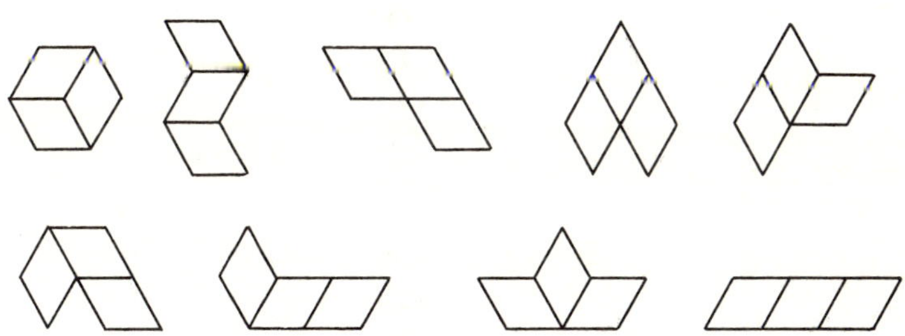

Fig. 26

Now see if the same can be done with another set of nine pieces coincidentally formed by joining two three-block trapezoids all possible ways.

Note also that the entire set of twelve size-six pieces might be assemblable into a 72-block rhombus or rhomboid many different ways. Which of those shown in Fig. 27 are possible to assemble?

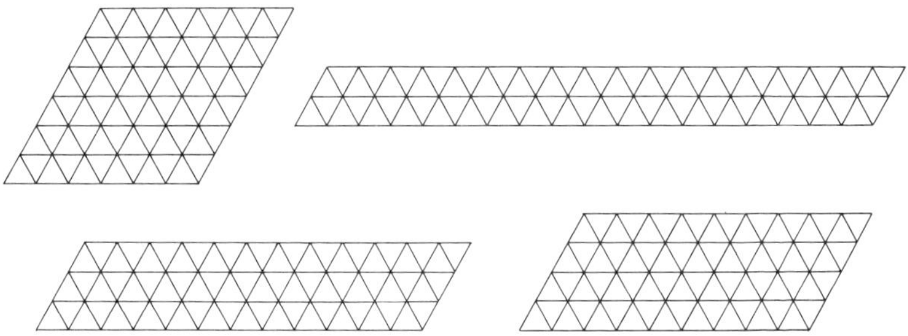

Fig. 27

For more information on amusements of this sort, see Martin Gardner's *Sixth Book of Mathematical Games from Scientific American.*

Squares as building blocks

Continuing in the same vein, we now consider squares as building blocks (Fig. 28), with the number of pieces summarized in Table 2.

Table 2

Size	Number of pieces	Total number of blocks
1	1	1
2	1	2
3	2	6
4	5	20
5	12	60
6	35	210

When we compare this summary with that for the triangle, the much greater versatility of the square as a combinatorial building block is apparent. The various pieces are popularly referred to as *polyominoes* after a book on the subject with that title by Solomon

Fig. 28

Golomb. The various sets of pieces have been the object of much investigation by mathematical analysts. Note that much of this analysis is concerned with proving mathematically where such pieces will or will not fit, which may or may not have much relevance to the design of practical geometrical puzzles.

Incidentally, the listing in Table 2 and others like it have been calculated for pieces of much larger size, often using a computer. Such pieces have little practical value in dissection puzzles. Beyond a certain size, the elegant simplicity of discrete dissection becomes obscured by complexity, going against the natural human inclination to reduce all things to their simplest and most functional common denominator. In combinatorial recreations of this sort, those that achieve their intended object using the fewer and simpler pieces are nearly always the more satisfactory.

The most obvious constructions for polyomino puzzle pieces are square or rectangular assemblies. If complete sets are being considered, then Table 2 suggests that only the size-four and size-five sets look interesting. First we will dispose of the size-four set. If the reader will mark and cut the five size-four pieces out of cardboard, it should be easy to convince oneself that it is impossible to assemble them into a 4×5 rectangle. But how can you be *sure*? This problem will be used as a simple example to illustrate two common analytical approaches to puzzles of this type.

Mark a 4×5 board on paper. Start with the straight piece and note that there are 13 different positions it can occupy on the board. But because of symmetries, only five are distinctly different. Of these five (shown in Fig. 29), three can immediately be eliminated by inspection. The remaining two can be analysed by placing a second piece, the square, in all possible positions and seeing what problems then arise. When one uses this method, a judicious choice of the first piece may save many steps. Usually a symmetrical piece is the best choice because there are fewer distinctly different ways that it can be placed.

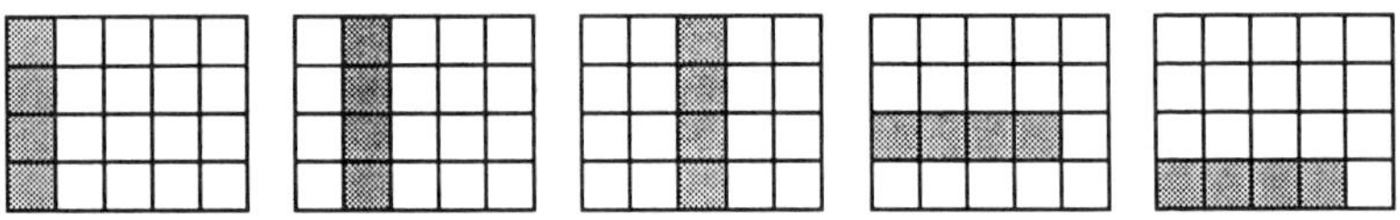

Fig. 29

An alternate technique frequently used for analysing problems of this sort is the following: Note that if the 4×5 board is checkered, it will always have 10 light and 10 dark squares. Now checker the pieces and note that four of them will always have two light and two dark squares, but the fifth will always have three of one colour and one of the other (Fig. 30). Consequently, they will never fit onto the board.

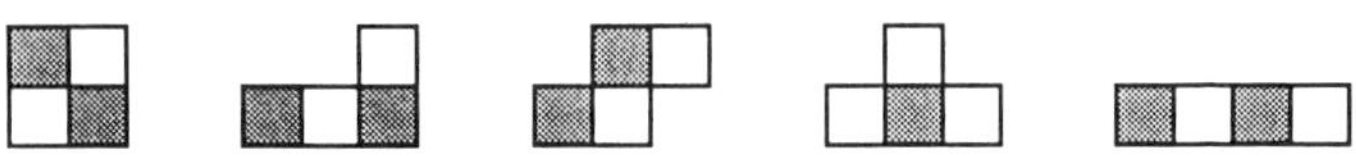

Fig. 30

Pentominoes

As already shown, joining five squares together all possible ways produces a set of 12 puzzle pieces, popularly known as *pentominoes* (Fig. 31). These occupy much of Golomb's book and have received much attention from others also. The total of 60 blocks is a most fortuitous number because it has so many factors. The earliest reference to a puzzle of this sort appears to be in *The Canterbury Puzzles*, by Henry Dudeney, published in 1907. The

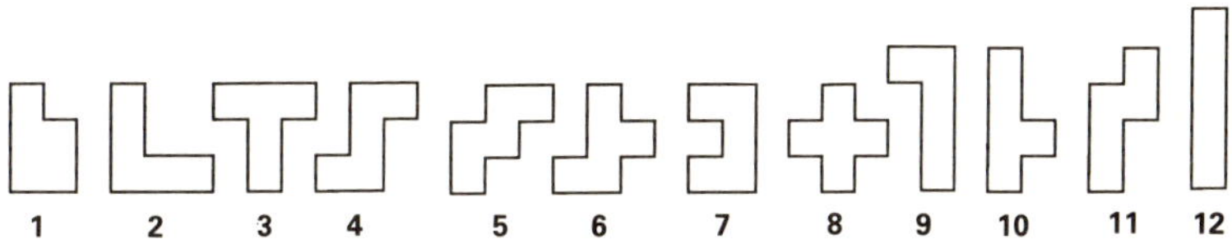

Fig. 31

idea is so obvious that it may have occurred to many persons independently.

The pentominoes are capable of being assembled into four different rectangles—3×20, 4×15, 5×12, and 6×10. The first investigation of these by computer was probably by Dana Scott in 1958. The results were summarized in an article by C. J. Bouwkamp in the *Journal of Combinatorial Theory* in 1969. There are 2, 368, 1010, and 2339 solutions to these four rectangular assemblies. One of each is illustrated in Fig. 32.

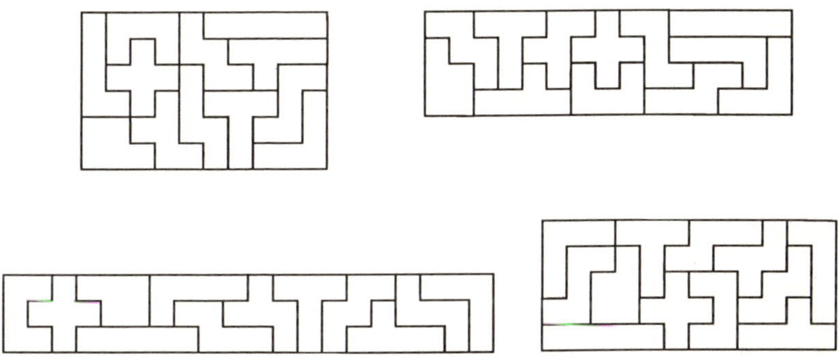

Fig. 32

With 2339 solutions, you might expect that placing the 12 pieces onto a 6×10 tray should be quite easy. If so, you are due for a surprise! One of the charms of puzzles of this sort is that the first few pieces fall into place and nestle together as though they were made just for each other's company. The next few pieces may be a bit more troublesome, but they finally settle down happily into place too. It is always the last one or two pieces that are the rascals. As you carefully rearrange things to suit them, then other pieces become the outcasts. Reluctantly you have no choice but to break up combinations that seemed so content together. Alas, you have made the situation worse instead of better, for now there are *three* that won't go in. In a moment of frustration, you are tempted to brusquely dump the lot out of the tray and start afresh. But no, you take the gentler and wiser approach of patiently switching just a few pieces back and forth, when suddenly the solution reveals itself as the remaining empty space just happens to match the last piece. As it drops snugly into place, there is a sense of resolution and harmony that any sensible person must welcome these days, especially if you have just scanned the headlines of the daily news or perhaps driven through Harvard Square in rush-hour traffic!

Although it was mentioned earlier that crude models will usually suffice for experimental work, that was not necessarily intended as a recommendation. Here is a case where one might well develop a deeper relationship with this captivating set of puzzle pieces by making them accurately and solidly of attractive hardwood, with a smooth finish and close fit and with a matching tray (Fig. 33). They will repay your consideration many times over.

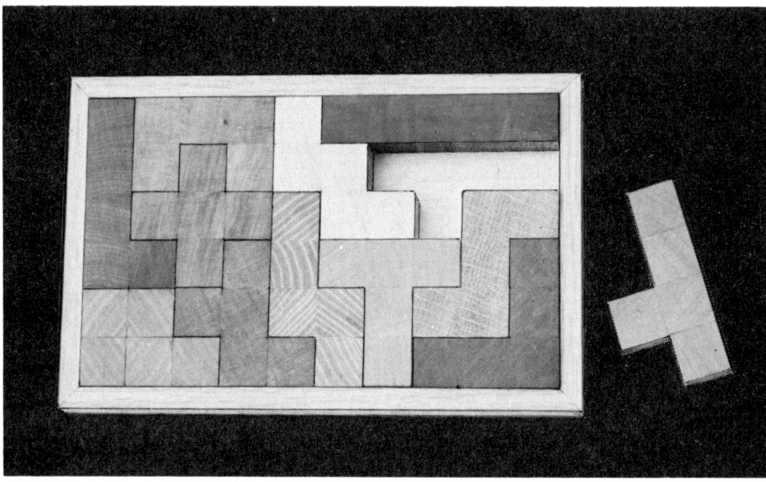

Fig. 33

What may at first seem like a random process of placing the first few pieces on the tray is anything but. Never underestimate the amazing power of the human brain, which gets even better with practice. For example, you will find some pieces much more co-operative than others. Piece no. 1 is the most tractable. Resist the temptation to place it early—it is your trump and should be kept in reserve until you really need it. Pieces that decline to fit nicely into the corners are the most troublesome. Piece no. 8 is the worst—it refuses altogether. Yet even it has a companion in piece no. 7, so let the pair of them mate. Try to fill the corners first, the ends next, and work toward the centre.

For an even more methodical (but less entertaining) approach, consider how a complete analysis of this puzzle might be made. Number the spaces on the 6 × 10 tray 1 to 60 as shown (Fig. 34). Always try to fill the lowest numbered unfilled space with the lowest numbered remaining piece. So, start by placing piece no. 1 on space no. 1. Since this piece has no symmetry, it can be oriented four different ways by rotation plus four more when flipped over, only six of which will cover space no. 1. With piece no. 1 in place, try placing piece no. 2 in the next numbered empty space. Piece no. 2 has four orientations by rotation, but because of symmetry it need not be flipped over (likewise pieces no. 3, no. 5, and no. 7). Continue placing pieces in this manner. Note that piece no. 4 has twofold rotational symmetry, so it has only two orientations plus two more when flipped. Piece no. 12 has both rotational and reflexive symmetry, so only two possible orientations. Piece no. 8 is the most symmetrical of all, with only one possible orientation. Furthermore, because of symmetries of the tray, the location of the starting piece can be confined to one quadrant.

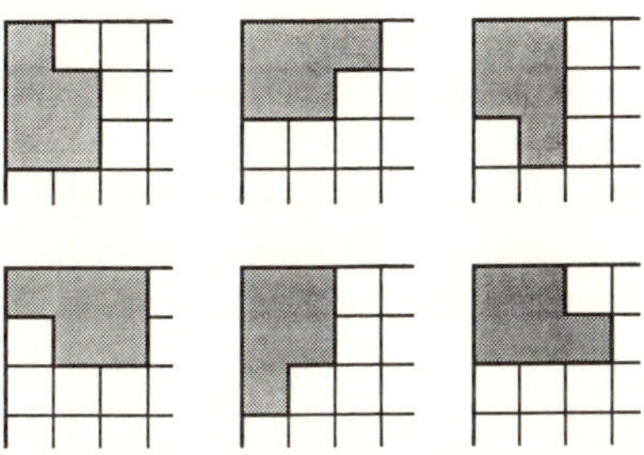

1	2	3	4	5	6	7	8	9	10
11	12	13	14	15	16	17	18	19	20
21	22	23	24	25	26	27	28	29	30
31	32	33	34	35	36	37	38	39	40
41	42	43	44	45	46	47	48	49	50
51	52	53	54	55	56	57	58	59	60

Fig. 34

Continuing methodically in this manner, one arrives at either a solution or an impasse. When an impasse is reached, the last piece placed is tried in every possible orientation. If that fails, the same is tried with the previously placed piece. Without belabouring all of the details, the point is that by proceeding methodically along these lines, or by some other similar scheme, one eventually tries every piece in every possible location and orientation, and compiles a complete list of solutions (or proves that none exists). If all of this sounds exceedingly arduous, it is indeed, and in the case of this particular example so much so as to be beyond practical human capability. This is where computers come into play. They are perfectly suited for this sort of mindless task. They do in seconds what might take a person days or months, and do so with much less likelihood of error.

More checkerboards

The joined-square combinatorial puzzles just described bear a close resemblance to the checkerboard dissections discussed in the previous chapter. The distinction between dissection and combinatorial puzzles has little to do with appearance, but rather with method of creation. The classification is not always precise, and the two categories tend to overlap. Consider the checkerboard puzzle shown in Fig. 35, which appeared in *The Canterbury Puzzles*. The pieces are printed on one side only so may not be flipped.

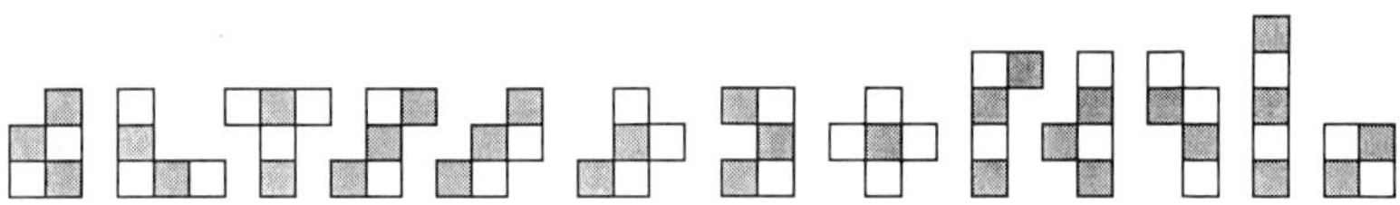

Fig. 35

At first glance, this might appear to be just another checkerboard dissection like those mentioned in the previous chapter. Upon closer scrutiny, however, it is obvious that the pieces were not created by a process of dissection. Rather, Dudeney must have taken the set of 12 pentominoes, added the 2×2 square to bring the square count up to 64, assembled all of the pieces into an 8×8 square, and lastly added the checkering.

In trying to solve this puzzle, one might start by assuming that Dudeney probably placed the square piece symmetrically in the centre for aesthetic reasons. (Before checkering, there are 65 solutions with this arrangement.) With the 2×2 checkered square thus centred, by placing the cross piece in each of its four possible locations, one discovers the impossibility of any such solution. This puzzle is known to have four solutions, but all with the square piece off-centre. Did Dudeney introduce this slight aesthetic anomaly just to confuse us? We will never know for sure, but if he did, why then would he have chosen a version with four solutions instead of just one, making it that much easier? A puzzle within a puzzle!

Of all the checkerboard puzzles in Slocum's *Compendium*, only one is said to be of recent vintage. It may appear at first glance to be just a variation of the Dudeney puzzle. But it was designed by Kathy Jones, which should alert any puzzle connoisseur to expect something thoughfully conceived and executed. The pieces are checkered on both sides and may or may not be the same on both sides. The puzzle has 1294 checkered solutions, and the 2 × 2 square can be in any possible position. It also solves several other problems. Three of the solutions with the 2 × 2 square in two different locations are shown in Fig. 36. Note that not quite enough information is given here to determine the exact checkering on both sides of all pieces. The puzzle is produced by Kadon Enterprises, Inc.

Fig. 36

The Cornucopia Puzzle

Solving or attempting to solve a mechanically manipulative puzzle analytically, with all of the action taking place unseen inside a computer, may not sound like much fun except for the computer fanatic. Be that as it may, the computer is now frequently being used as an aid in the design and optimization of combinatorial puzzles. Many examples will be given in this book. One of the more impressive of these is the Cornucopia Puzzle.

It was shown on page 27 that joining six squares all possible ways produces a set of 35 puzzle pieces. Now, from this set, eliminate all pieces having reflexive or rotational symmetry and all those containing a 2 × 2 square, because they are less desirable for various reasons already explained. The remaining 17 pieces are the set of *Cornucopia* pieces (Fig. 37).

A subset of any 10 of these pieces will fit onto an 8 × 8 board leaving four empty squares. There are 10 different ways that these four empty squares can be arranged in fourfold symmetry (disregarding reflections) as shown in Fig. 38.

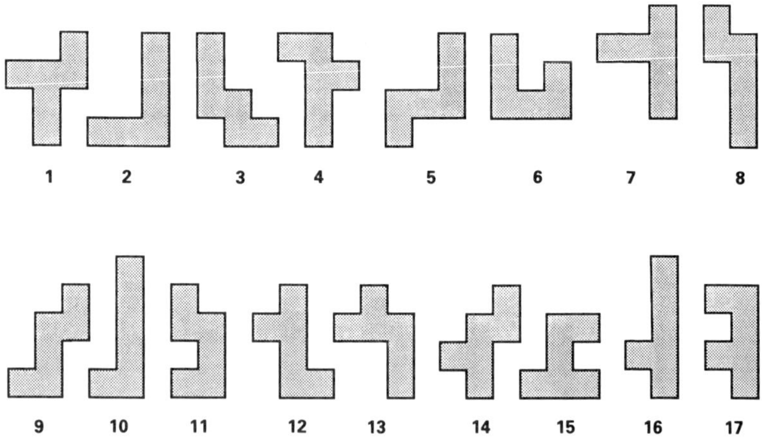

Fig. 37

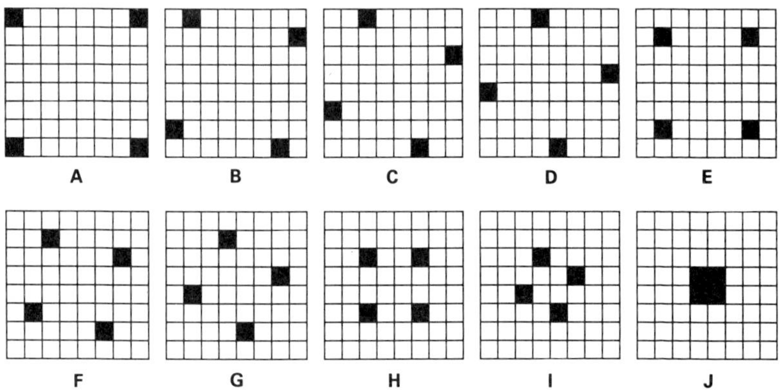

Fig. 38

A subset of 10 Cornucopia pieces can also be assembled to form a solid 6×10, 5×12, or 4×15 rectangle. (Note: the 3×20 rectangle is impossible. Can the reader discover a neat and simple proof of this? Hint: place piece no. 1, 3, 4, 5, 7, 9, 12, 13, or 15 anywhere in the 3×20 rectangle and see what problem then arises.)

Combinatorial theory shows that a subset of 10 pieces can be chosen from a set of 17 pieces 19 448 different ways. Which of these subsets will fit any of the boards shown in Fig. 38, and in how many different ways? Expert puzzle analyst Mike Beeler decided to find the answers to these questions with the aid of a computer. Even with state-of-the-art equipment and clever short cuts, this

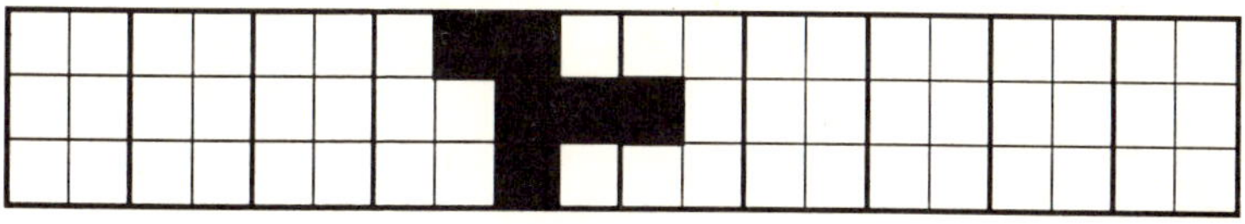

Fig. 39

probably involved more computation than any previous puzzle analysis, and by a wide margin. The final results show 8203 usable subsets and 105 902 solutions, any one of which constitutes an interesting and challenging puzzle problem, hence the name *Cornucopia.* This suggested the possibility of producing a series of Cornucopia puzzles whereby each set of pieces would be unique, and each with its own unique solution or solutions. (The idea itself is also believed to be unique!) One hundred such sets were produced in wood in 1985 and are now in the hands of puzzle collectors. The remainder are contained in a stack of papers a foot high!

At the beginning of the Cornucopia project, as the computer started to spew out solutions, we wondered if any subset would be found that made all 13 patterns. Preliminary results indicated this to be very unlikely. To our surprise, however, near the end of the run one prolific subset was discovered that failed to do so by the narrowest margin. It is shown in Fig. 40, with a tabulation of all its copious solutions given in Table 3. To gain some appreciation of the power and speed of a computer, the reader is invited to make this subset of pieces and try to solve just one of the other 56 solutions listed. Now imagine all 57 of them being solved in a few seconds!

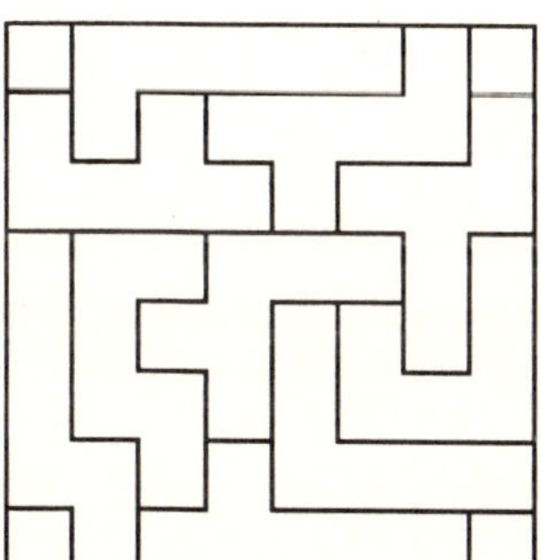

Fig. 40 The copious cornucopia

Table 3

Pattern	Solutions
A	7 (one shown)
B	1
C	0
D	1
E	11
F	1
G	1
H	1
I	4
J	1
6×10	15
5×12	12
4×15	2

With polyomino-type puzzles like Cornucopia, when many solutions are known, here is an interesting exercise: display all of them together and have several friends judge which they consider to be the most and least pleasing to the eye. If there is any consistency in the judging, try to determine what characteristics are common to those judged most or least pleasing. Finally, try to define these characteristics mathematically.

After staring at thousands of Cornucopia solutions, the author has selected the two shown in Fig. 41 as being a good pair for illustrating this game. Everyone polled by the author preferred the same one. The other one has at least four easily recognizable and describable flaws. What are they? (Answer later in book.) Paradoxically, perhaps the most distinguishing feature of a pleasing polyomino pattern is its lack of any distinguishing features! Evidently, the mind's eye prefers randomness in such designs. We

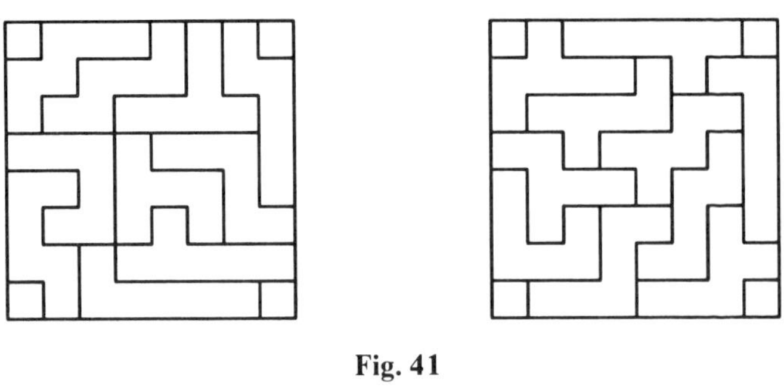

Fig. 41

all know what randomness is, or *think* we know, until we try to define it mathematically. Randomness is, almost by definition, something that cannot be defined mathematically! And even if the rules for randomness could be stated mathematically, what about the rules for the rules?

To further compound this strange paradox, at the same time the eye goes to the opposite extreme and automatically takes for granted absolute conformity to the square grid as an unspoken rule. Any deviation from this, as in the example in Fig. 42, cries out as blatantly as would a sour note in a Mozart serenade or an obscenity in an Emily Dickinson poem!

Fig. 42

It is interesting to note that the basic element of a Cornucopia-type puzzle is symmetrical—a square, and the overall assembly is also symmetrical—likewise a square. A dissymmetry occurs between these two extremes in the permutated shape of the puzzle pieces. Thus, the order—*symmetry—dissymmetry—symmetry*—represents in itself another, more abstract sort of symmetry (Fig. 43a).

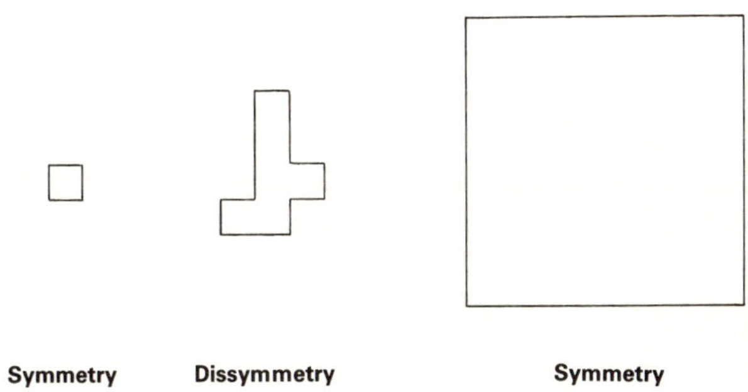

Fig. 43a

A typical die-cut jigsaw puzzle is an example of a different order of symmetry, which is itself non-symmetrical (Fig. 43b).

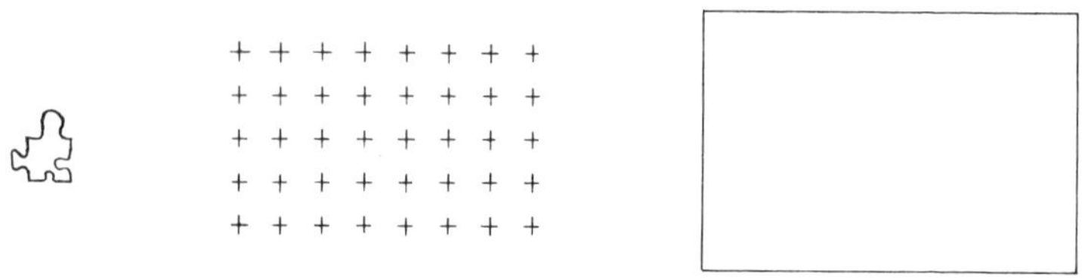

Fig. 43b

A third order of symmetry would be represented by bathroom tiles on a typical floor (Fig. 43c).

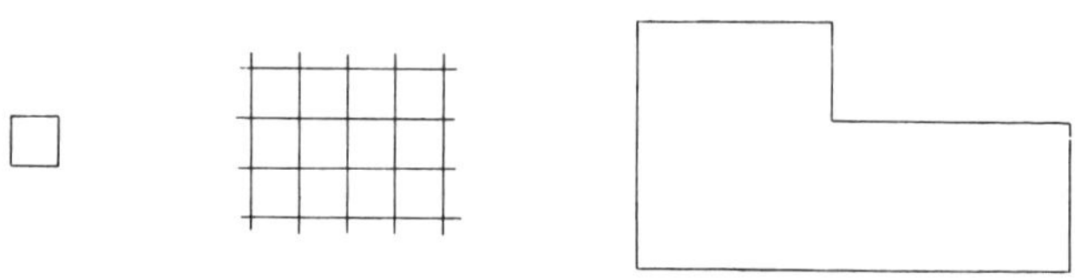

Fig. 43c

Can you discover other orders of symmetry? This term, and in fact the whole concept of *order of symmetry*, was developed just as a curiosity as this chapter was being cut and pasted for at least the tenth time. This might be an interesting subject of study in itself. Practically all puzzles described in this book are inherently of the symmetrical order: *symmetry—dissymmetry—symmetry*. The intriguing symmetries of the polyhedral shapes are often what attract the attention of the viewing public, much to the chagrin of the designer, for all of the creative work lies hidden within and is so often overlooked.

Traditionally, artistic endeavours from music to poetry to oil paintings have nearly always been framed in symmetry of some sort, or at least were until this century. Yet symmetry is a hopelessly sterile artistic medium within which to actually work. All creativity involves a judicious departure from symmetry inside this symmetrical framework.

Hexagons as building blocks

Shown in Fig. 44 are all the ways that hexagonal blocks can be joined, up to size-four. (Incidentally, should the curious reader wish to construct a set of hexagon pieces of size-five, one tedious but sure way to do this is to add an extra block to all of the size-four pieces in every possible position and then throw out the duplicates. You should end up with 22 pieces.)

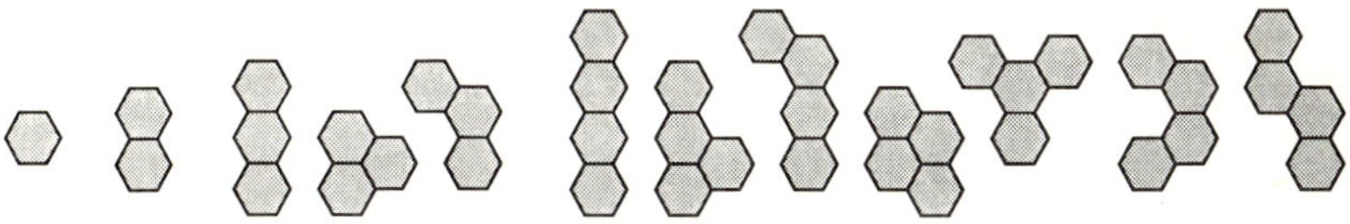

Fig. 44

Table 4

Size	Number of pieces	Total number of blocks
1	1	1
2	1	2
3	3	9
4	7	28

The most obvious problem shapes to construct with such pieces are hexagonal clusters. These are shown in Fig. 45 in increasing size.

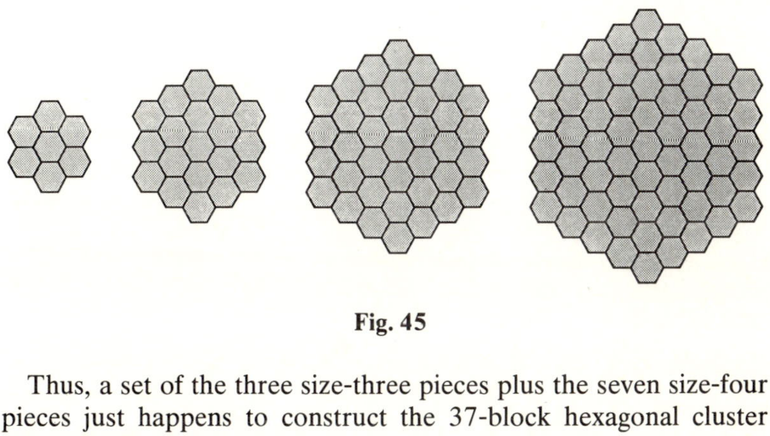

Fig. 45

Thus, a set of the three size-three pieces plus the seven size-four pieces just happens to construct the 37-block hexagonal cluster (Fig. 46). It also constructs a snowflake-shaped figure plus many

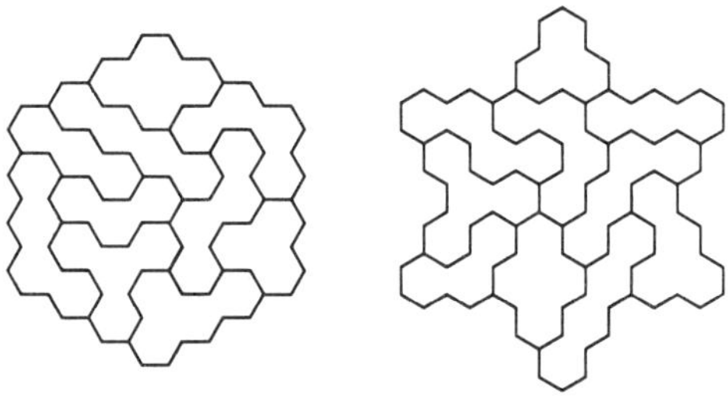

Fig. 46

other geometrical and animated shapes. Beeler found by computer analysis that the hexagonal cluster has 12 290 solutions, and the snowflake pattern (from which a commercial version of this puzzle derived its name) has 167 solutions.

One of the special charms of this set of pieces is that it lends itself so well to creating geometrical, artistic, and animated puzzle problems. Just a few examples taken from the 10-page instruction booklet that came with the Snowflake Puzzle are shown in Fig. 47. The others are left for the reader to rediscover or improve upon. By the way, this is but one more example of a recreation in which young children excel. Many of the design patterns in the Snowflake Puzzle booklet were created by children under ten years of age.

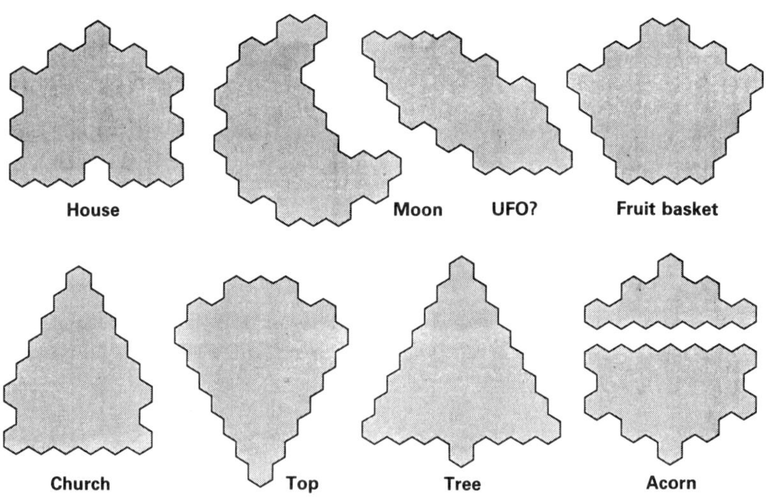

Fig. 47

Chapter 3

Cubic block puzzles

Given the popularity of puzzles made up of squares joined together in different ways, it does not require too much imagination to realize that cubic blocks might be joined together in similar fashion to make three-dimensional puzzles. When measured in terms of the number of different assembly combinations possible for a given set of pieces, the cube must be the ultimate combinatorial building block. Add to that the fact that the pieces are easy to make, to visualize, to describe and to illustrate. They also pack nicely into a box or rest on a flat surface. No wonder they are so popular!

The 3×3×3 cube

The earliest reference to $3 \times 3 \times 3$ cubic block puzzles may be one shown in the classic *Puzzles Old and New*, by Professor Hoffmann (Angelo Lewis), published in London in 1893, and not to be confused with the recent Botermans and Slocum book of the same name. It shows a puzzle called the *Diabolical Cube*, which is rather a misnomer as it is one of the easier puzzles of its type. The six pieces, illustrated in Fig. 48, assemble into a $3 \times 3 \times 3$ cube 13 different ways. Since all of the pieces in this puzzle have reflexive symmetry, it necessarily follows that every solution must either be self-reflexive or be one of a reflexive pair. It is customary not to count these reflexive pairs as two different solutions. This particular version of what has now become a very common type of puzzle is unusual in that all of the pieces are flat and contain different numbers of cubes increasing in arithmetic progression.

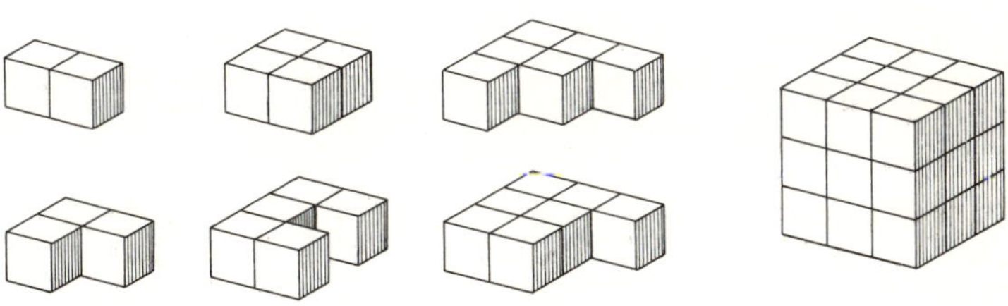

Fig. 48

The next reference known to the author for the 3 × 3 × 3 cube is a version that appeared in *Mathematical Snapshots*, by Hugo Steinhaus, published by Oxford University Press in 1950. Puzzle historians might well be puzzled by this half-century gap. With all of the interest in burrs, etc. during that time, could there have been no interest in cubic blocks? The version in the Steinhaus book (Fig. 49) has two solutions that are slight variations of each other and of medium difficulty. It is referred to as *Mikusiński's Cube* after its originator, the Polish mathematician J. G. Mikusiński.

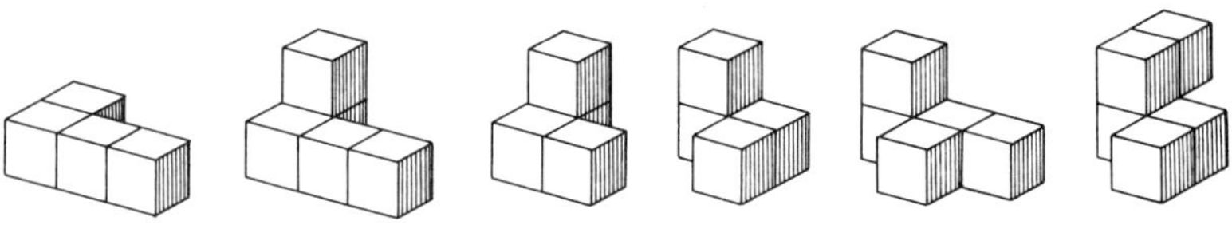

Fig. 49

Nearly everyone must be familiar with Piet Hein's seven-piece Soma Cube (Fig. 50), which is said to have been invented around 1936 and which enjoyed great popularity and commercial success around the 1960s. With 240 possible solutions, the 3 × 3 × 3 assembly is almost trivially simple. Its popularity may have been due more to the well-conceived instruction booklet showing many different problems and pastimes possible with the set.

The popularity of Soma lingers to this day. Sivy Farhi publishes a booklet containing over 2000 problem figures. There have been versions with colour-matching problems, with number problems on the faces, and so on.

Variations on the 3 × 3 × 3 cube that have been published within the last two decades are now too numerous to mention. Commercially successful puzzles nearly always spawn a host of imitations. Even if some are well conceived or even an improvement over the original, they are almost certain to languish in obscurity,

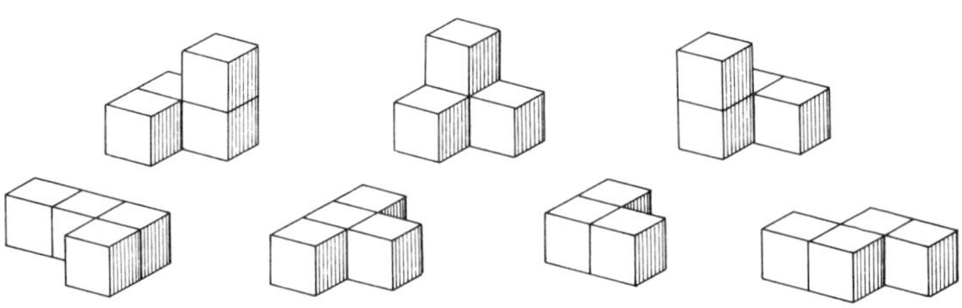

Fig. 50

since puzzle fads tend to run in cycles with no mercy on come-lately look-alikes. But we need not be concerned with that here. As an archetype, the $3 \times 3 \times 3$ cube is a superb combinatorial puzzle—simple in principle and embodiment, yet with many secret charms still lying buried inside. Perhaps we can dig a few of them out.

With puzzles of this type, there is an optimum number of pieces; and as you tinker with them, you soon gain an intuitive sense of what that number is. There is no way that a four-piece version can be very difficult, although the one shown in Fig. 51 does have the intriguing property of being serially interlocking, meaning that it can be assembled in one order only. Is a five-piece serially interlocking version possible?

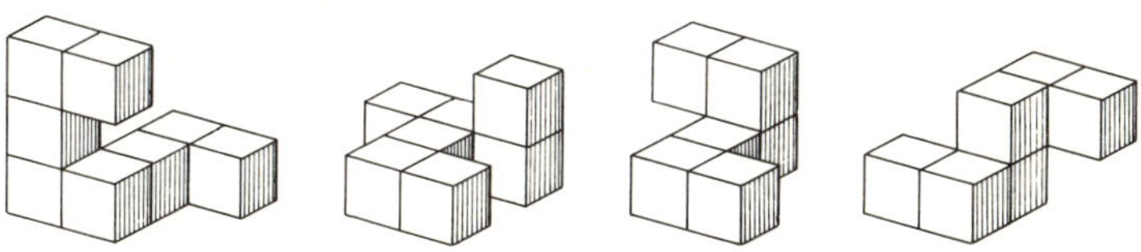

Fig. 51

The five-piece and six-piece versions of the $3 \times 3 \times 3$ cube are the most interesting. Some of the five-piece designs are surprisingly confusing. The six-piece designs have the added advantage that they usually can be assembled into many other symmetrical problem shapes. (A very cleverly designed five-piece puzzle might have this feature, too.) In order to make a systematic study of this puzzle family, the first step is to list all ways that four or five cubes can be joined (as shown in Fig. 52).

The six-piece version of the $3 \times 3 \times 3$ cube will be considered first. For aesthetic reasons, one might prefer that all the pieces be the same size, but this is impossible, so the nearest approximation is to use three four-block pieces and three five-block pieces. It is also desirable that all pieces be non-symmetrical, but this is likewise impossible, so two of the four-block pieces will have an axis of symmetry. All pieces will of course be dissimilar. Of the several thousand such combinations possible, the author tried several that proved to have either multiple solutions or no solution, until finally finding one with a unique solution. It is shown in Fig. 53. It was produced at one time as the Half Hour Puzzle.

Although it was intended to construct only the $3 \times 3 \times 3$ cube, Hans Havermann and David Barge have discovered hundreds of other symmetrical constructions possible with this set of puzzle

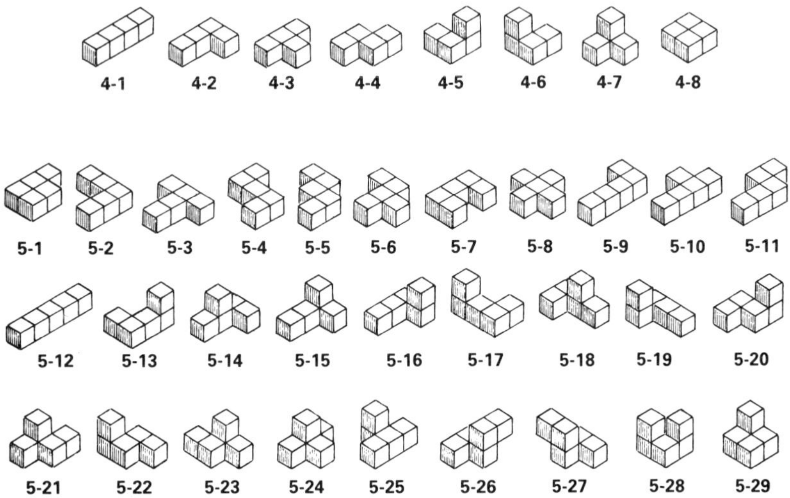

Fig. 52

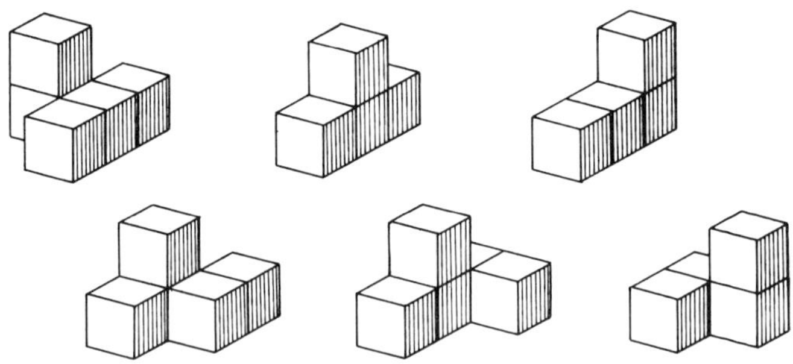

Fig. 53

pieces, a few of which appear in Fig. 54. All of these figures have at least one axis or plane of symmetry, and they represent most but not all of the types of symmetry possible with this set. The cube has 13 axes and 9 planes of symmetry. Two of the figures have one axis and two planes of symmetry. Another has one axis and one plane. All the others have one plane of symmetry only. Challenge: with this set, discover a construction with one axis and four planes of symmetry—i.e. the same symmetry as a square pyramid. One is known. Are there more?

In the five-piece versions of the $3 \times 3 \times 3$ cube, there may be three five-block pieces and two six-block pieces, and none need be symmetrical. The number of such possible designs must be in the

Fig. 54

thousands, and many of them are surprisingly difficult. One is shown in Fig. 55, but readers are encouraged to experiment with original designs of their own, not necessarily using the guide-lines suggested above.

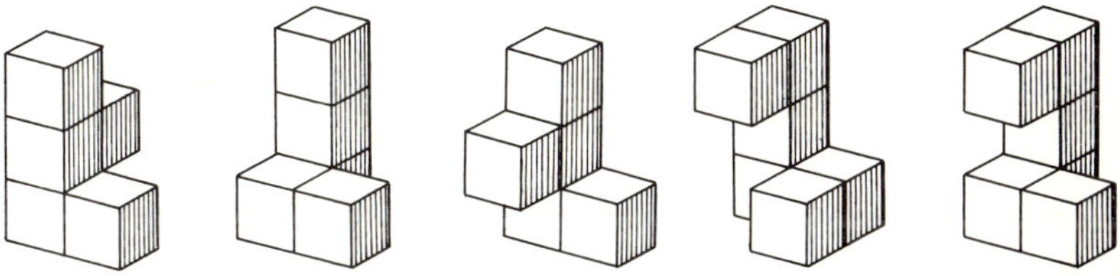

Fig. 55

Throughout this book, and throughout the world of geometrical puzzles in general, it is taken for granted that the sought-for solution is not only symmetrical, but usually the *most* symmetrical possible shape—in this case, the cube. When multiple problem shapes are considered, highest priority is given to those having the most symmetry. Evidently, one of the most basic and deeply rooted instincts of mankind is an eye for symmetry, whether in the arts, the sciences, or whatever. Trying to give reasons for so ingrained an instinct is perhaps a risky business, but here is an attempt so far as puzzles are concerned:

For reasons already explained, ideally the solution of a combinatorial puzzle, by definition, begins with the individual pieces in a state of greatest possible disorder, meaning all dissimilar and non-symmetrical. A symmetrical solution, then, goes to the opposite extreme, and does so against the natural tendency in the world toward disorder and randomness. Only the human brain is capable of doing this. Practically every human endeavour involves at least some attempt to make order out of disorder, but nowhere more graphically than in the symmetrical solution of a geometrical dissection puzzle. It is the one point to which all paths lead upward and from which one can go no higher. To put it another way, the object of a well-conceived geometrical recreation is usually obvious enough so as to require minimal instructions. One tends to associate complicated instructions with unpleasant tasks—the definitive example being of course the filing of income taxes. Contrarily, life's more enjoyable pastimes tend to require no instructions at all!

Polycube pieces fit together so naturally that some persons find recreation in simply assembling random 'artistic' shapes and thinking up imaginative names for them. When they don't resemble anything, the tendency is to call them 'architectural designs'. (Does this tell us something about the present state of architectural design, or at least the public's perception of it?)

The solid tetrominoes

Note that four cubes can be joined eight different ways. Packing a set of these pieces into a $4 \times 4 \times 2$ box makes a neat but quite easy puzzle. There are said to be 1390 possible solutions. They also pack into a $2 \times 2 \times 8$ box, and can be split into two $2 \times 2 \times 4$ sub-assemblies.

The solid pentominoes

Another popular cubic block puzzle is the set of 12 solid pentominoes. These are of course the set of pieces made by joining five squares all possible ways, discussed on page 28, except in this case cubic blocks are used in place of squares. The idea of a puzzle set made of such pieces is so obvious that it probably occurred to several persons independently. The earliest references known to the author are associated with Martin Gardner's mathematical recreations column in *Scientific American* around 1958. They were implied in an article by Golomb in *The American Mathematical*

Monthly, December 1954, and are discussed in his book *Polyominoes*.

The solid pentominoes (Fig. 56) pack into the following rectangular solids: $2 \times 3 \times 10$, $2 \times 5 \times 6$, and $3 \times 4 \times 5$. Bouwkamp's computer analysis found there to be 12, 264, and 3940 solutions to these, and these numbers have been confirmed by many other analysts. If you did not learn your lesson with the flat pentominoes, and you think that, with 3940 solutions, packing these pieces into a $3 \times 4 \times 5$ box ought to be easy, you are in for an even bigger surprise this time!

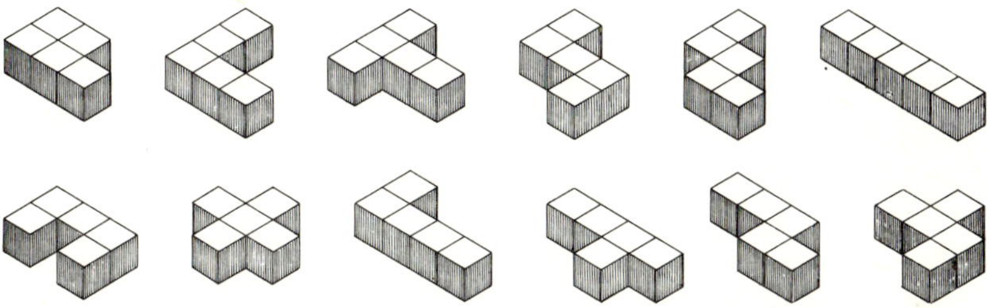

Fig. 56

The solid pentominoes make a very satisfactory set of puzzle pieces when accurately crafted of nice hardwoods and packaged in a suitable box. Many interesting puzzle problems and pastimes, using either the full set of pieces or subsets, are crammed into *Quintillions*, a booklet published by Kadon Enterprises. Or you can invent your own puzzle problems. Fig. 57 shows a set where each piece is of a different kind of wood in contrasting natural colours.

A checkered pentacube puzzle

There are 12 pentacubes that are flat (the solid pentominoes) and 17 that are not. There are 12 that have an axis of symmetry and 17 that do not. There are 12 that neither lie flat nor have an axis of symmetry. If we arbitrarily eliminate the two of these that fit inside a $2 \times 2 \times 2$ box, then a set of 10 pieces remains (see Fig. 58).

According to a computer analysis by Beeler, these pieces pack into a $5 \times 5 \times 2$ box 19 264 different ways, and it is not very difficult to find one of them. To make this puzzle more interesting, the pieces are checkered (Fig. 59). There are two ways that one

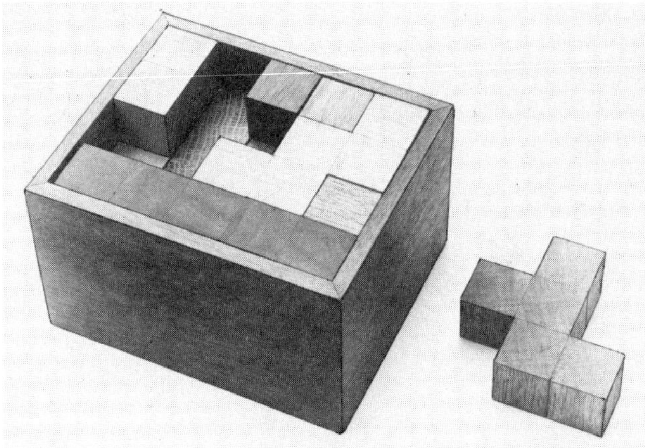

Fig. 57

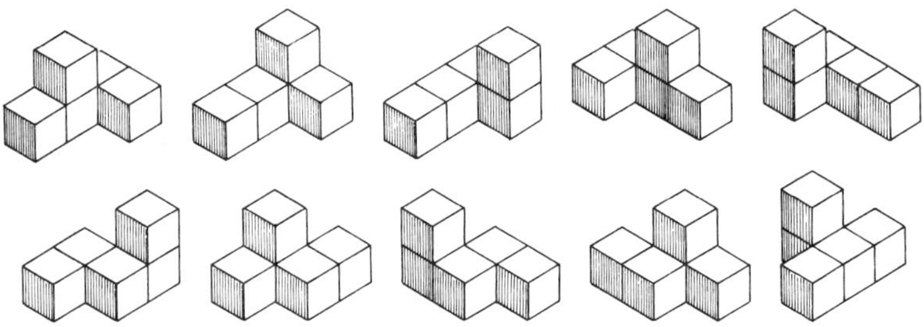

Fig. 58

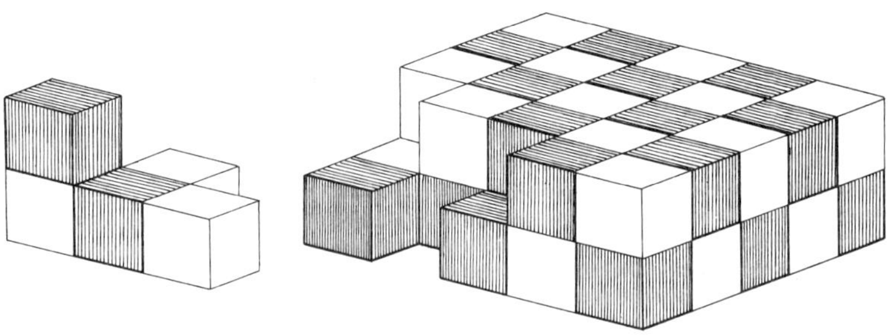

Fig. 59

might go about this. You could randomly checker the pieces and then try to assemble checkered solutions. There are 512 different ways of checkering the pieces, of which 511 have solutions and one does not. So it would be remotely possible, if you were exceedingly unlucky, to end up that way with an impossible puzzle. The better way is to assemble the puzzle first and then add the checkering. This way you are sure of having a solution. Now try to find a second perfectly checkered solution with this set of pieces. Of the 511 ways of checkering the pieces with solutions, 510 of them have multiple solutions and one is a unique solution. So there is this very slight chance that your puzzle may not have a second checkered solution, but you may never know for sure, because finding the other solutions is very difficult (unless you use a computer). How remarkable that out of the 512 possible checkerings, just *one* should be impossible to assemble checkered and just *one other* should have a unique solution!

Polycubes in general

Puzzle pieces made up of cubes joined different ways (polycubes) are of course unlimited in size and infinite in number. Those of size-six are called *hexacubes*, size-seven *heptacubes*, and so on. Questions such as how many there are of each size would more likely be pursued as curiosities in mathematical analysis rather than for practical puzzle applications. The most satisfactory polycube-type puzzles are those using small-sized pieces in seemingly simple constructions.

The interesting design possibilities for polycube-type puzzles are practically limitless. Furthermore, the pieces are among the easiest to make. For those with no access to woodworking tools, cubic wooden blocks can be obtained from educational supply stores. Also from this same source, plastic cubes that snap together are handy for experimental work.

In addition to the booklets already mentioned, *World Game Review* serves as a clearing house for new ideas in polycubes and related recreations.

Rectangular blocks

Closely related to the polycube puzzles are the so-called 'packing problems' using rectangular blocks. Again, many of these are of interest primarily to mathematical analysts, but some of them also make satisfactory assembly puzzles. Take for example the *Slothouber-Graatsma Puzzle*, which calls for three $1 \times 1 \times 1$ cubes and six $1 \times 2 \times 2$ blocks to be packed into a $3 \times 3 \times 3$ box. There is only one solution. Another is known as *Conway's Puzzle* after its inventor, mathematician John Conway. It calls for packing three $1 \times 1 \times 3$ blocks, one $1 \times 2 \times 2$ block, one $2 \times 2 \times 2$ block, and thirteen $1 \times 2 \times 4$ blocks into a $5 \times 5 \times 5$ box. It is quite difficult unless one happens to be an expert in this particular branch of mathematics.

An interesting puzzle is suggested by joining $1 \times 2 \times 2$ blocks in pairs all possible ways. The resulting 10 pieces are shown in Fig. 60. They can be assembled into a $4 \times 4 \times 5$ solid, and there are said to be 25 solutions. Now eliminate the two pieces that are themselves rectangular, and see if the remaining eight (shaded) will assemble into a $4 \times 4 \times 4$ cube. After you have become convinced that they will not, find a set that will by duplicating one piece and eliminating one piece, and note the interesting pattern of symmetry in the solution.

In the same vein, a simple puzzle project is to find all the ways that $1 \times 1 \times 2$ blocks can be joined in pairs. Then assemble them into a rectangular solid and discover one solution having a pattern of reflexive symmetry.

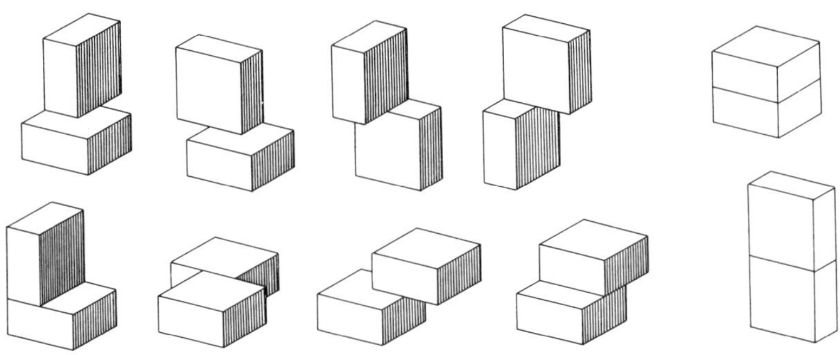

Fig. 60

Chapter 4

Interlocking block puzzles

All of the puzzles described thus far (except one) have been non-interlocking. Most of them employ a tray or box to hold the pieces in place. The puzzles to be described in this chapter, and throughout the rest of the book, are *interlocking*. In other words, they hold themselves together. To be more precise, an interlocking puzzle is here defined as one in which the last step of assembly (or first step of disassembly) necessarily involves the sliding of mating surfaces parallel to each other. Such puzzles tend not to come apart without deliberate effort. A box is no longer needed to hold them, so they can be any geometrical shape and can be displayed in full view when assembled. There is more freedom in manipulation of the pieces. Beyond these obvious practical advantages, isn't there something intrinsically more satisfying in things that stay together rather than fall apart by themselves? (Anyone who owns a car like mine will understand!)

Cubic block puzzles

The polycube pieces in the previous chapter were formed by joining cubic blocks together different ways. None of the pieces thus formed up to size-five are sufficiently crooked to have much practical use as interlocking puzzle pieces. More importantly, the combinatorial approach does not lend itself very well to the design of interlocking block puzzles.

The most obvious method of designing an interlocking cubic block puzzle is to start with the complete pile of blocks, held loosely together by your imagination or some other means, and remove one piece at a time. A $4 \times 4 \times 4$ cubic pile is a good size for this, with its millions of possible dissections. Depending upon just what the objectives are, quite a bit of experimenting may be required to achieve the desired results. Again, the plastic play blocks that snap together are handy.

A commonly accepted rule for combinatorial puzzle design is that the pieces all be dissimilar and non-symmetrical. The fundamentals of good design also require that the simplest possible pieces be used that will do the job. Given the $4 \times 4 \times 4$ cube then, this translates into maximizing the number of pieces. What is the maximum number of dissimilar non-symmetrical pieces that will assemble into an interlocking $4 \times 4 \times 4$ cube? (Answer unknown.)

The Convolution Puzzle

Because of the millions of possible ways of dissecting the $4 \times 4 \times 4$ cube into dissimilar, non-symmetrical, interlocking puzzle pieces, additional aesthetic considerations may be introduced to make the design process more of an art rather than just a series of random choices. The puzzle could be made serially interlocking, meaning that it can be assembled in one order only. Also, by using $1 \times 1 \times 2$ blocks in the construction, symmetrical patterns can be realized on the six outside faces. Shown in Figs. 61a and 61b is a seven-piece dissection that attempts to achieve all of these. It was once produced as the Convolution Puzzle. It has one surprising step in assembly that requires a rotation, which is not possible unless certain edges are rounded ever so slightly. Can any reader devise a way to correct this mechanically slight but mathematically crippling deformity in an otherwise satisfactory design?

The Three-Piece Block Puzzle

Challenge: join just 10 cubic blocks together to make three puzzle pieces that interlock to form a puzzle having threefold axial symmetry. Impossible? Of course, if you assume that the blocks are joined face-to-face. But when cubic blocks are joined by their half-faces or quarter-faces, many new possibilities arise, as well as hopeless confusion!

All of the information required to construct such a puzzle is contained in the drawings in Fig. 62. This is such an amazing puzzle, it would be a shame to spoil it by giving the solution here. But note the following: interlocking puzzles of this sort must be quite accurately made to be entirely satisfactory or even to be assemblable at all. Usually the easiest way to achieve this is to glue at least some of the joints with the blocks held together in their assembled positions. Since that option is not given here, unless the reader is able to achieve the difficult feat of solving this puzzle on paper, the alternative is to first make a rough model using soft material or rubber cement. Then, after the solution is discovered, a more accurate model can be made of hardwood.

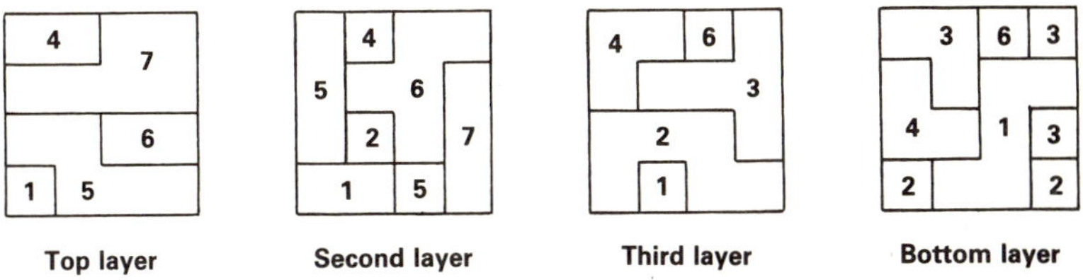

Fig. 61a

Fig. 61b

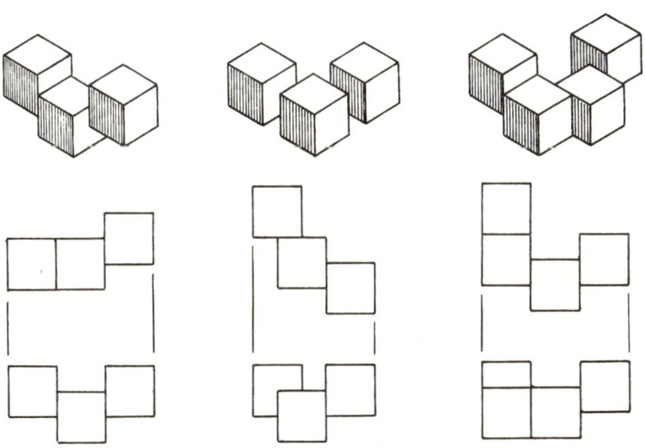

Fig. 62

This puzzle has an interesting history. The one symmetrical face of the assembled puzzle happens to resemble a certain corporate logo. The company wanted a simple puzzle incorporating this pattern for some sort of promotional scheme. So the arrangement of six of the blocks was already determined. All that was required to complete the design was the addition of four more blocks in a sort of triangular pyramid and a judicious choice of glue joints to make it into an interesting interlocking puzzle. So the company got what they wanted—except for one thing. It turned out to be anything but simple! Do not be discouraged if you cannot solve it straight away—it has baffled experts!

Chapter 5

The six-piece burr

Puzzles consisting of interlocking assemblies of notched sticks are often referred to as *burr* puzzles, probably from being pointed or spur-like in assembled appearance. By far the most familiar of all burr puzzles, and probably of three-dimensional puzzles in general, is the so-called six-piece burr. Its origin is unknown, but it has been traced back to at least 1803 in Germany, where it appears in a catalogue of G. H. Bestelmeier. Some persons know it as the *Chinese Puzzle* or *Chinese Cross*, probably because it has been mass-produced in the Orient since the early 1900s, but there does not appear to be any evidence that the idea originated there.

David Bruce has put forth the plausible conjecture that some of the earliest puzzles may have been but slight modifications of practical objects. For example, note the familiar interlocking box shown in Fig. 63, consisting of six notched boards. Did some whimsical box-maker decide to have fun one day in his spare time, or did he perhaps just run out of nails? Whatever, this may well have been the origin of the six-piece burr. With six identical pieces, as suggested by the illustration, it is clearly impossible to assemble. There are several obvious ways to modify one or more of the pieces

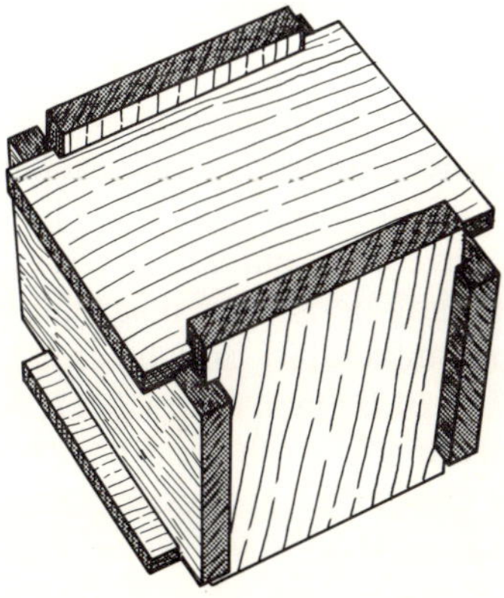

Fig. 63

to make it assemblable, and a good exercise for the amateur puzzle-maker is to see how many of these ways he or she can discover. With a penny slot, it becomes a toy bank—a good first puzzle for any youngster.

The standard six-piece burr consists of six notched square sticks of arbitrary equal length not less than three times their width, arranged symmetrically in three mutually perpendicular intersecting pairs. If the square cross-section of the sticks has a dimension of two units, then all notches are one unit deep and one unit wide or some exact multiple. To put it another way, all notches can be regarded as being made by removal of discrete cubic units, or to put it still another way, all pieces can be regarded as being built of cubic units. All of the notches are made within the region of intersection with the other sticks, so that when the puzzle is assembled no notches show and it has apparent symmetry (Fig. 64).

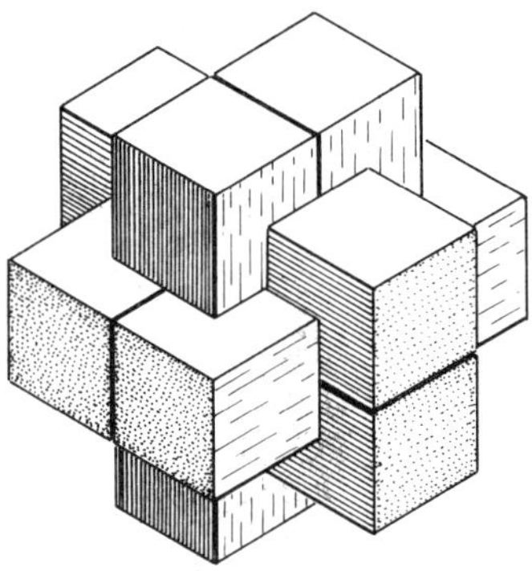

Fig. 64

The six-piece burr is actually a large family of designs, since the designer has a wide choice of how to notch each of the pieces. Over the years, variations of the six-piece burr have received much attention from puzzle inventors and authors. Directions for making them can be found in many books and magazines. Several different versions have been manufactured and patented. The earliest US patent is No. 1 225 760 of Brown, dated 1917, with several others following shortly thereafter. Most toy and novelty stores have a few burr puzzles on their shelves or in their catalogues. Traditionally, these have been uninspired time-worn versions with sliding key and internal symmetries. Consequently this fine puzzle has suffered a chronically tarnished image. To make matters worse, over the years many inventors have tinkered with bizarre embellishments to give the basic burr puzzle their own stamp of identity. The patent files reveal many such ill-conceived contraptions, including those with strings and holes, hidden pins, rotating keys, and other secret locking devices. Evidently taking their cue from certain composers of modern 'music', they have thrown in odd intervals, incongruously sharpened or flattened pieces, confusingly large numbers of parts in hopeless disharmony with each other, and other jarring complications. Within the last decade or two, though, the six-piece burr has emerged from this decadent period to become once again the quintessential interlocking puzzle, thanks largely to the work of Bill Cutler.

There are twelve cubic units in each piece that are candidates for removal (Fig. 65). Removing these in all possible permutations would theoretically result in 4096 different pieces, but by discarding symmetries and those that cut the pieces in two, the number of *practical* pieces is 837. No one knows how many different ways sets of six such pieces can be combined and assembled, but they number in the millions. Cutler limited his analysis to only those combinations that make a solid assembly with no internal voids. Using a computer, he found there are 369 *usable* pieces and they can be assembled into a solid burr 119 979 different ways. These results were published in the *Journal of Recreational Mathematics*, Vol. 12, No. 2, 1979–80, and were summarized by Martin Gardner in *Scientific American*, January 1978.

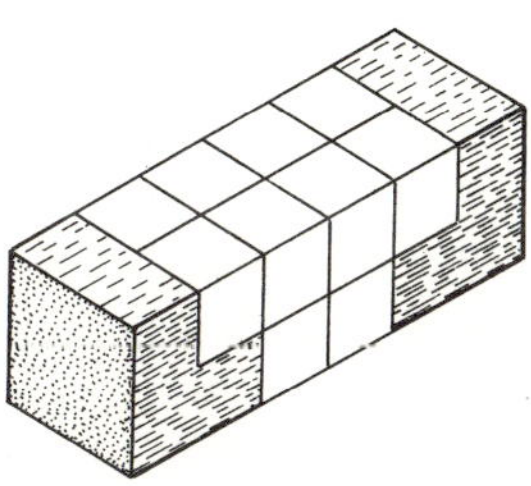

Fig. 65

The burr pieces can be divided into two groups—those with simple notches that can be milled out directly with a saw or dado blade (*notchable* pieces), and those with blind corners and edges that must be chiselled out or made by gluing in cubic blocks (*unnotchable* pieces). The notchable pieces are the more desirable from both practical and aesthetic considerations. Some puzzle analysts have limited their investigations to notchable pieces, of which there are 59 (including the one with no notches). It has been customary to consider only solid assemblies, in which case there are only 25 usable notchable pieces, and these are commonly referred to as the *set of 25 notchable pieces.* They can be chosen in sets of six and assembled solid in 314 different ways. Some of this was calculated independently by several different analysts, with or without a computer. All of it has been confirmed and organized by Cutler.

In Fig. 66, piece *A* is notchable. Piece *B* is not notchable. Piece *C* can be made with a saw but cannot be assembled with other notchable pieces without producing voids, so it is not included in the set of 25 notchable pieces, which are shown in Fig. 67.

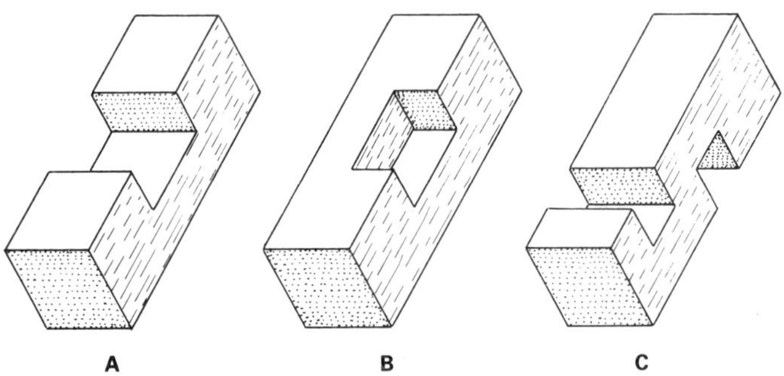

Fig. 66

Burr No. 305

Cutler's computer analysis told only what was possible, not what was most interesting. Actually it might possibly have done that too if appropriately instructed. For example, from the list of the 314 solid notchable combinations, suppose one first eliminates all those using duplicates (or triplicates) of identical pieces and pieces having an axis of symmetry. Also eliminate combinations with more than one solution. This narrows the field down to 18, of which all but one (and its mirror image) employ a rather common and uninteresting two-piece key arrangement. What emerges from this screening process is a marvellous burr. It is called *Burr No. 305* because of its location in Cutler's tabulation. It uses pieces 6, 12, 14, 21, 22, and 23 (Fig. 68).

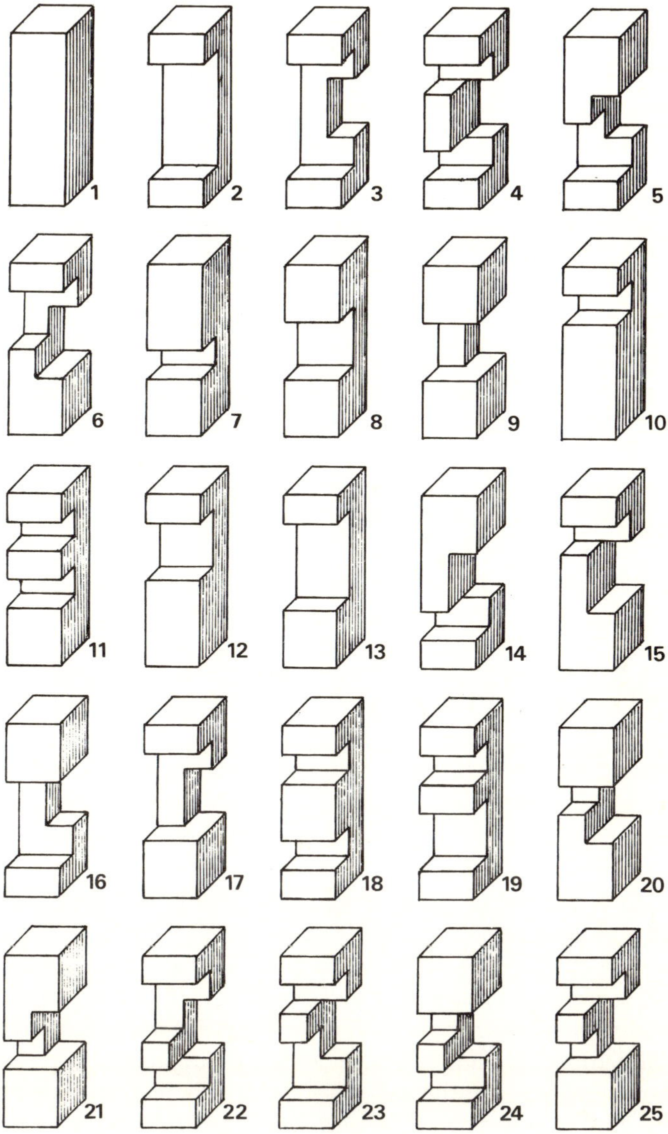

Fig. 67

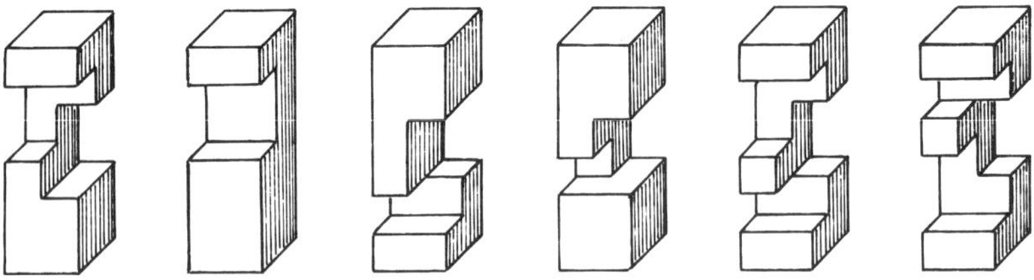

Fig. 68

Difficulty Index and Burr No. 306

This is an appropriate point at which to digress for a moment and introduce the idea of a *difficulty index* for a combinatorial puzzle. Puzzles must by definition have some element of difficulty. Making a puzzle more difficult may in some circumstances be an improvement in design, if not carried to extremes and if not to the detriment of other considerations. In any case, some way of predicting the relative difficulty of similar puzzle designs would be a useful tool for the designer.

Consider the solid six-piece burr. Given a drawing of the assembled burr or some familiarity with it, the only real problem is determining the relative location and orientation of the six pieces. Select any one of the six pieces at random for the bottom piece. Usually it is obvious from the notching which side should face the centre. Now for the back piece, one has a choice of any of the remaining five, and it can be turned end-for-end, hence a total of ten possibilities. For the next piece, say on the left, there are six choices, and so on. Thus, to make a complete analysis of the puzzle by trying every piece in every position, there are a total of $10 \times 8 \times 6 \times 4 \times 2$ or 3840 possibilities to be considered. This number divided by the number of solutions is the *difficulty index* of that particular design.

The difficulty index of Burr No. 305 is 3840. While that may seem like a large number of moves, most of them are skipped by using common sense, and so this would be a puzzle of medium difficulty. Identical pairs of pieces, symmetrical pieces, and multiple solutions all decrease the difficulty index. There is one charming type of piece known as an *ambiguous* piece, because you cannot tell from the notches which side should face the centre, and there are different degrees of ambiguity. Piece No. 9 in Fig. 67 is an example of the most ambiguous type because any one of its four sides might face the centre. This would increase the difficulty index by an additional factor of four, but because it is also symmetrical the net increase would be a factor of two.

The mischievous role of the ambiguous piece was not taken into account in the analysis that led to the illumination of Burr No. 305. Adding this newfound ingredient to the recipe, another delectable puzzle comes to light—Burr No. 306, illustrated in Fig. 69. It uses pieces 6, 9, 12, 21, 22, and 23, and has a difficulty index of 7680.

Note that a set of seven pieces will allow both Burr No. 305 and Burr No. 306 to be constructed.

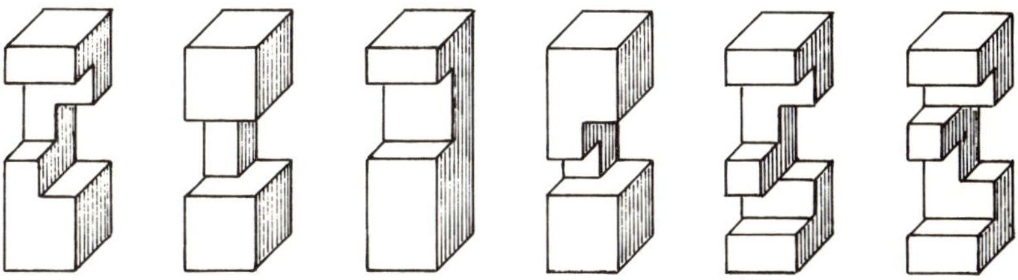

Fig. 69

Higher-level burrs and Bill's Baffling Burr

So far we have discussed only burrs with no internal voids. Historically, solid burrs have received the most attention. No satisfactory explanation has ever been given for this, but perhaps it is simply the notion that many things in life tend to be more satisfying when they are solid—building foundations, financial investments, friendships, and so on. Also, Bill Cutler points out that only by limiting his program to solid burrs was the analysis practical, as otherwise the computation time would have been too long. The recent flurry of activity in designing ever more entertaining (meaning, to some, fiendishly difficult) burrs has shifted attention to burrs that do not come directly apart (or go directly together), but rather involve the shifting back and forth of pieces, or groups of pieces, within the partially assembled burr. Some of these are so baffling as to discourage a professional locksmith, yet they are basically just standard burrs using the 837 practical pieces. They all necessarily have one or more internal voids.

One of the best of this new breed of burrs is Bill's Baffling Burr, designed of course by Cutler. It uses two unnotchable pieces, both of which are easily made from notchable pieces by gluing in one and two extra blocks. It has seven internal voids. This is an unusually large number of voids for a burr with only one solution, and contributes to its difficulty, for there are 24 *apparent* solutions but only one that is possible to assemble. Thus you may think you have found the solution and are wondering how to get the last piece in place when most likely you have stumbled upon one of the 23 false solutions. It was stated earlier that the pieces could be of arbitrary length. With some of these more complicated burrs, this is no longer true. Bill's Baffling Burr (Fig. 70) cannot be assembled if the pieces are longer than three times their width.

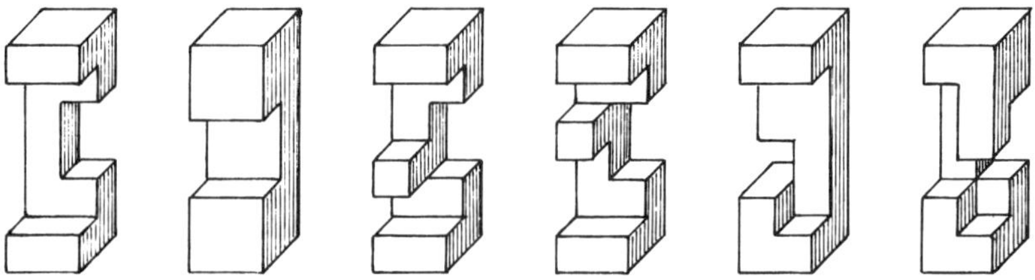

Fig. 70

Bill's Baffling Burr is referred to as a *level-five* burr, meaning that five separate shifts are required to release the first piece. This new yardstick of devilry has spurred some rivalry among puzzle experts to see who can come up with even higher level burrs. Around 1985, Gaby Games of Israel came out with an amazing level-seven-four burr, meaning that seven moves are required to release the first piece and four more are required to release the second piece. Would someone next discover a level-eight burr? Well, not exactly. Recently, Peter Marineau surprised the puzzle world with a level-nine burr! Is this the upper limit? Probably not. The Marineau burr, shown in Fig. 71, achieves its remarkable stunt with surprisingly simple pieces. Two are identical, another two are a reflexive pair, and another one is self-reflexive.

Perhaps the reader will now be encouraged to wander off into this vast wilderness of hidden notches and explore some of them further. For the puzzle connoisseur, a well-crafted six-piece burr is

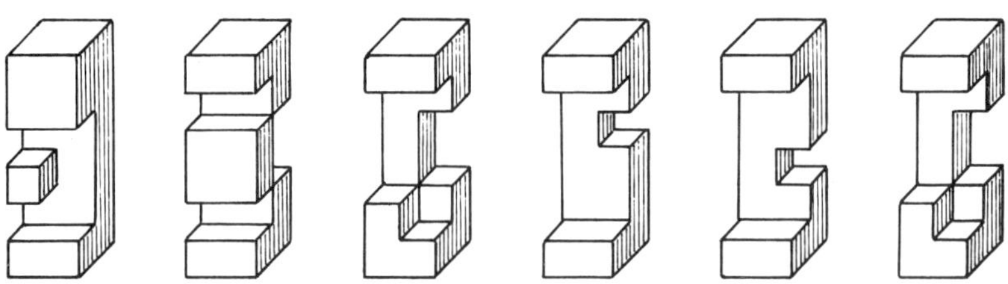

Fig. 71

the embodiment of good design—simple, direct, and eminently functional! For the hobbyist, the burr is well suited for a workshop project, and helpful woodworking tips are given later in this book. In particular, the would-be puzzle inventor will find much to explore beneath the deceptively familiar exterior of the six-piece burr.

Considering the large number of possible assemblable sets of the 837 practical pieces, recently estimated by Cutler to almost certainly exceed one billion, any one of them chosen at random is likely to be a new and original, but totally uninspired, design. The first step, then, is to decide just what features one considers most desirable. A few guide-lines have been suggested here, but there may be other, better ideas that have been overlooked. Originality, psychology, and aesthetics all play a role at this stage of the creative process. The second step is seeking the combination that best achieves one's goal, and this is essentially an analytical and mechanical problem.

Imagine a computer being programmed to methodically print out all of the probably billions of assemblable standard six-piece burrs. All but a handful would be new and original designs—or would they? Does merely being different constitute originality? There is a curious musical analogy. With conventional discrete musical notation, one could, in theory at least, program a computer to print out every possible musical theme, given enough time and unlimited supply of paper. Buried within this mountain of papers would be all of the most sublime works of the great masters of the past and of those perhaps to come in the next Renaissance. But then how could they be found from amongst the random noise? The whole exercise would amount to nothing.

Trying to improve upon an existing burr design can be an enlightening exercise. For example, as a maker of puzzles, one is always trying to reduce the number of unnotchable pieces. Moving or removing just one offending unit block seems innocent enough, but it nearly always causes havoc. Attempts to correct the problem just create more problems. (Sounds familiar?) Sometimes you work through a loop of changes and end up back where you started. It is slow work, for every new change requires an analysis of all possible solutions. Some analysts use a computer for this. It does in seconds what otherwise might take hours or even days. Others not in such a rush may enjoy the mental exercise in traditional methods of analysis, using pencil and paper. For them, the analysis *is* the puzzle, so why not relax and enjoy it?

Chapter 6

Larger (and smaller) burrs

The family of burr or notched-stick puzzles is a large and prolific one, with offspring numbering in the hundreds or even thousands depending upon how one counts them, and with more being born all the time. This book is not intended to be a compendium of puzzle inventions, past and present. One yardstick for inclusion is the extent to which the underlying idea behind the puzzle is logical and mathematical rather than simply mechanical. Symmetry is an important consideration. In this chapter we will conclude the discussion of square-stick burrs by considering only those having certain kinds of symmetry. This is an appropriate point at which to discuss symmetry.

Symmetry

The term *isometric symmetry* was introduced on page 2 without explaining what it meant. A three-dimensional object is said to have isometric symmetry if it has identical non-coplanar axes of symmetry. In other words, it exists in a sort of geometrical vertigo, with no identifiable 'upright' orientation, no top or bottom, front or back, left or right, all being the same. All of the Platonic solids have this property, including their various truncated and stellated variants. Rectangular solids (except the cube) and pyramids (except the tetrahedron) do *not* have it. The three-dimensional object in question can be anything from a polyhedral solid to a cluster of solids, a nesting of sticks, or whatever.

There is another sort of symmetry that most of the burr puzzles in this book have, sometimes referred to as *homogeneity* or *congruence.* It is illustrated by the two drawings in Fig. 72. On the left is the standard six-piece burr. The 12-piece burr on the right is representative of a popular family of puzzles, sometimes called *pagodas,* which lack homogeneity because not only are the sticks of various lengths but also their relative positions are distinguishable.

The term *homogeneous isometric symmetry* is so awkward that it will not be used throughout this book, but rather will be implied. Most well-conceived burr puzzles have it, and the lacking of it must

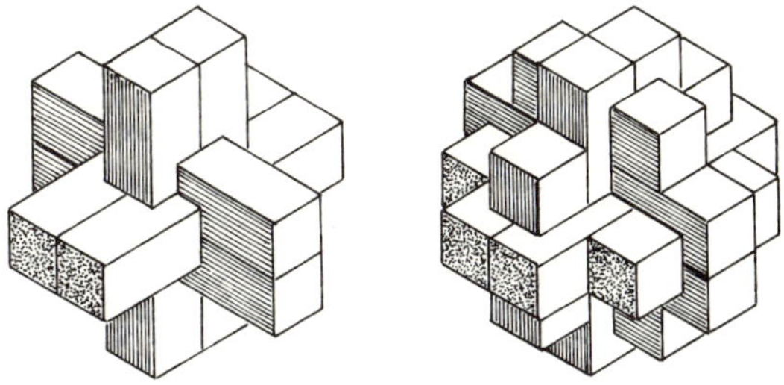

Fig. 72

be considered an aesthetic blemish. Like so many aesthetic considerations, this one too is rooted firmly in practicality. Most interlocking puzzles have a key piece or sliding axis that constitutes the first step of disassembly. In a symmetrical burr, all pieces have equal standing and are indistinguishable from one another when assembled, thus coyly hiding their identity beneath a geometrical masquerade.

A distinction is made between *apparent* symmetry and *total* symmetry. When a puzzle has apparent symmetry, as do practically all well-designed geometrical puzzles, the assembled external shape is symmetrical but not necessarily the insides. When a puzzle has total symmetry, all of the internal surfaces of dissection are symmetrical as well. Such puzzles necessarily have all pieces identical, limiting their possibilities for combinatorial problems, but there are some intriguing exceptions involving colour symmetry.

For both practical and aesthetic reasons, as already discussed, anyone who tinkers with geometrical puzzles usually takes for granted the concept of symmetrical external form and internal dissymmetries without giving it much thought. It is interesting to note that all higher animals also have this property, although the exact reasons for it are not at all obvious. If one's body and brain were entirely bilaterally symmetrical, could one tell the difference between left and right, throw a ball, or use a typewriter? Could one even think, in the usual sense of the term?

The three-piece burr problem

Symmetrical rectilinear burrs can be made of 3, 6, 12, or 24 sticks—no other sizes are possible. The basic six-piece burr was discussed in the previous chapter. The most obvious form for the 12-piece burr is that shown below, left. Notice that the axes of the intersecting sticks are not offset as in the six-piece burr, but instead intersect with each other. In order to understand what problem this creates, consider the simple three-piece burr shown in Fig. 73 on the right.

A little reflection should convince the reader that with any sort of conventional rectilinear notching, the three-piece burr is impossible to assemble, or even more obvious, impossible to *dis*assemble. When you see a three-piece burr of this type, you can be sure that either one of the notches has been rounded so that a piece rotates, or else the sticks have diagonal or otherwise unconventional notches. The same applies to the 12-piece version.

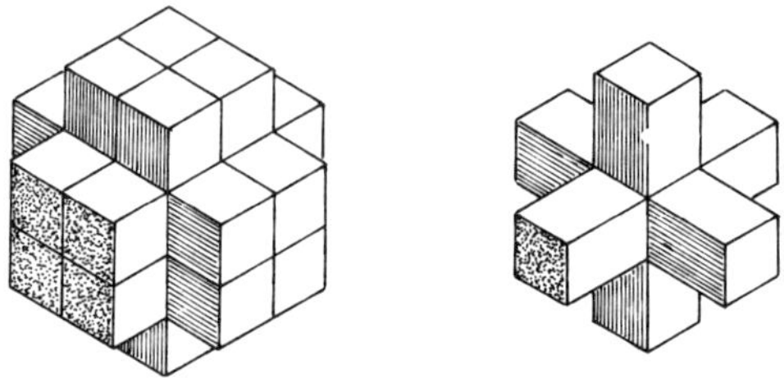

Fig. 73

Practical 12-piece burrs

One way of overcoming the problem just explained is to space the sticks apart in the 12-piece burr, as shown in Fig. 74 on the left. Symmetry is maintained. The possibilities for notching combinations are virtually limitless. There is a well-known variation of this with the sticks spaced farther apart so that it resembles a cage (Fig. 74, centre). Sometimes a ball is placed inside. There is yet another variation, but non-homogeneous, with three more pairs of sticks added to fill the centre spaces, even more complicated to design, to solve, or even to explain (Fig. 74, right).

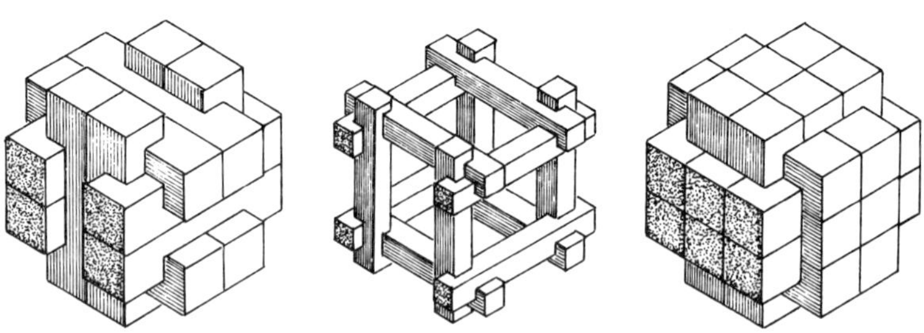

Fig. 74

The Altekruse Puzzle

There is one other symmetrical 12-piece burr that is a classic and quite unlike any of the others mentioned thus far. It is shown assembled in Fig. 75, together with one of its 12 pieces, all of which are identical. US Patent No. 430 502 was granted to William Altekruse in 1890 for this puzzle. The puzzle has been popular for a long time and manufactured in many different forms with many different names (except Altekruse). The Altekruse family is of Austrian-German origin. Curiously, the name means 'old cross' in German, which has led some authors to incorrectly assume that it was a pseudonym. A William Altekruse who is presumed to be the grantee of the patent came to America as a young man in 1844 with his three brothers to escape being drafted into the German army. Could he have brought at least the germ of the idea with him? Whatever the case, it is a most interesting burr. By the way, note that if one insists on being precise, it is not quite symmetrical visually because the asymmetrical notch arrangements reveal themselves.

Fig. 75

The Altekruse Puzzle, sometimes known as the 12-piece burr, has an unusual mechanical action in the first step of disassembly by which two halves move in opposition to each other. This may come as quite a surprise to those accustomed to the more familiar burr types with key piece or pieces. Depending upon how it is assembled, this action can take place along one, two, or all three axes independently but not simultaneously. If two extra pieces are available, there is a surprising 14-piece solution.

Variations of the Altekruse Puzzle

The interesting variations of this puzzle are quite numerous, and probably others await discovery. In the standard Altekruse Puzzle, each piece has three notches, with the two end notches facing in the same direction. There is a variation in which some pieces have notches facing in opposite directions, and such pieces can be either one of a reflexive pair, as illustrated in Fig. 76. Which combinations using such pieces are possible?

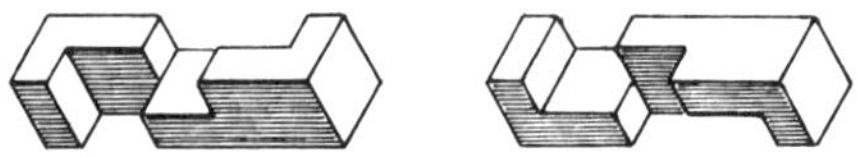

Fig. 76

The repetitive structure of Altekruse pieces can be extended indefinitely to create larger puzzles. Before considering these, note the diminutive version shown in Fig. 77 that uses six pieces of two notches each—three right-handed pieces and three left-handed. Try to solve this puzzle visually, and then discover an interesting variation that does not use equal numbers of right-handed and left-handed pieces (and do not forget that it must be assemblable).

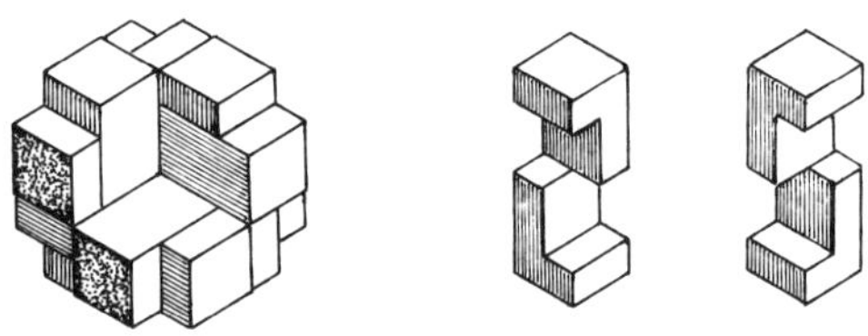

Fig. 77

There is a version that uses 24 sticks, four notches in each, 12 right-handed and 12 left-handed (Fig. 78, left). There is a version that uses 36 or 38 identical sticks of five notches each (Fig. 78, right), and so on *ad infinitum*. There are rectangular versions in even greater number. Note that none of these larger versions is homogeneous. Once the basic principle is understood, these larger versions are not very difficult to assemble except that some trial and error may be required to figure out the correct order of assembly. They also require more dexterity than some of the others, and it helps if the pieces are accurately made.

Another interesting variation of the Altekruse Puzzle uses pins and holes in place of notches. In its simplest version, each piece has one pin and one hole, with six right-handed pieces and six left-handed pieces. An unusual feature of this version is that, with a large supply of pieces to work with, they can be connected end-to-

Fig. 78

end to make longer sticks and larger, more complex assemblies without limit. To make things more interesting, there need not be equal numbers of the two types of pieces, and there may also be pieces with pins facing in opposite directions. For even more entertainment, add pins or holes in the centres of the pieces. Just figuring out all the possible pieces is quite a task, and analysing all of the puzzling 12-piece assemblies should keep someone occupied for a long time.

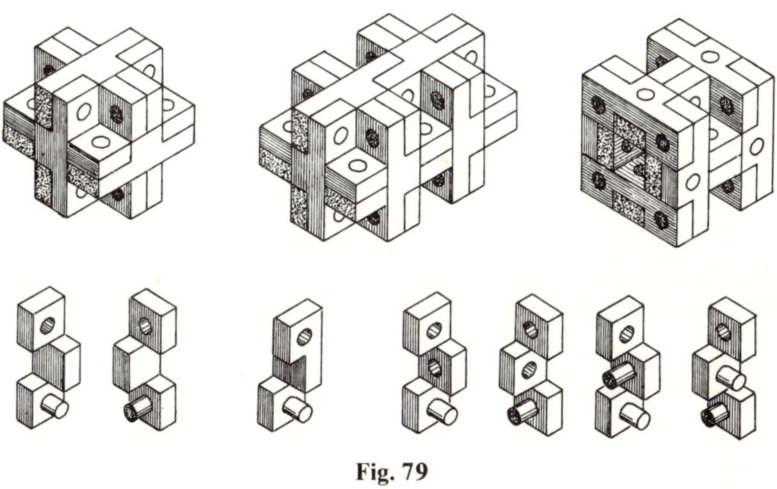

Fig. 79

The Pin-Hole Puzzle

Like the design described above, this one also uses pins and holes. The basic puzzle consists of six 1 × 1 × 3 bars and six dowels of length 3. Each bar has three holes slightly larger than the dowels. The puzzle pieces are fabricated as shown in Fig. 80, using brads to hold the dowels in place. One version of the puzzle uses one pin, one bar, two crosses, and three elbows. This puzzle can be assembled one way only but is quite easy.

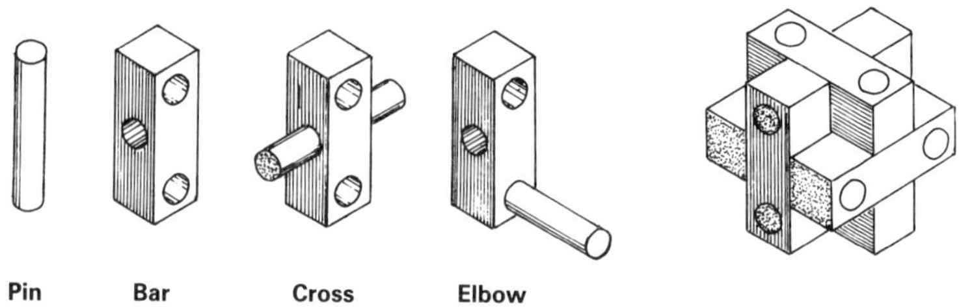

Fig. 80

By having larger sets of pieces and including one more type of piece twice as long, many larger and more complicated figures can be constructed, two examples of which are shown in Fig. 81.

Fig. 81

The Corner Block Puzzle

The Pin-Hole Puzzle has an interesting variation. Eight cubic blocks are added to the corners of the Pin-Hole Puzzle, making the assembled shape cubic (Fig. 82a).

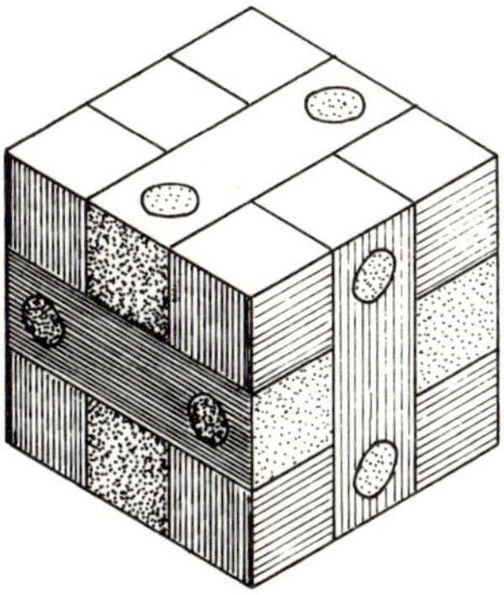

Fig. 82a

Each cubic block might be attached to any one of the three bars against which it rests. Thus the puzzle designer faces a choice of 3^8 or 6561 different ways of attaching the blocks. Another way of looking at the problem is to consider all the different types of pieces that could result. When one considers all the ways that one or two blocks can be added to the three basic pieces (bar, cross, elbow), there are 18 possible augmented pieces. Six of these are less desirable because they have an axis of symmetry, leaving the 12 pieces shown in Fig. 82b. Problem: from this set of 12 pieces, find a subset of six pieces that assembles one way only. The author has tinkered with this problem off and on for years without success. For some reason not understood, the solutions always seem to occur in pairs or more. To simplify the problem somewhat, note that piece 1 must always be used, plus three more pieces with single blocks and two with double blocks. Thus, there are 150 possible subsets.

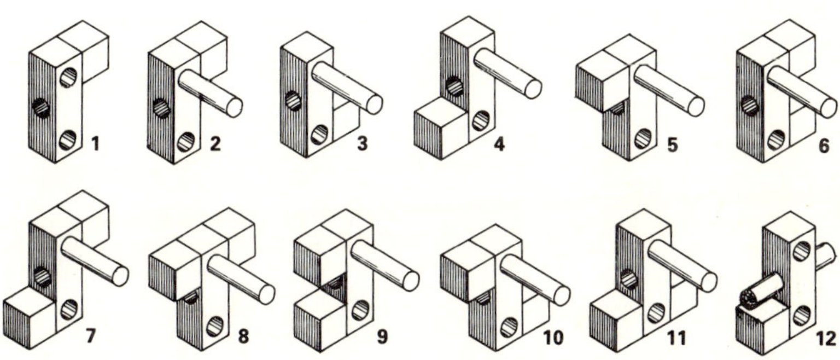

Fig. 82b

Here is one fairly satisfactory combination with a pair of solutions: pieces 1, 2, 3, 7, 8, and 12. Can the reader improve upon this? The Pin-Hole Puzzle and Corner Block Puzzle are not really burrs. They sneaked into this chapter as close relatives. This theme is carried forward in chapters 13 and 22.

A 24-piece burr

The number of practical ways that 24 notched sticks are symmetrically assemblable is very limited. Only one is known to the author, shown in Fig. 83. It uses 23 identical pieces and one key piece having an extra notch. With an illustration to follow, assembly of the puzzle is mostly a test of dexterity. There is also a surprising solution that uses 24 identical pieces without any key piece, and it requires even more dexterity.

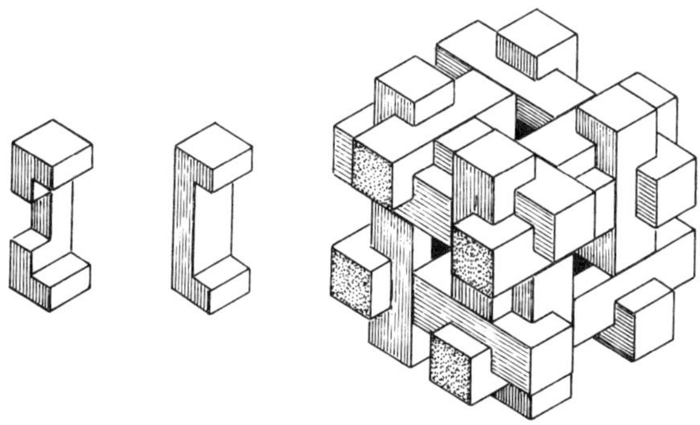

Fig. 83

The really interesting feature of this puzzle set is that other interlocking assemblies are possible using fewer pieces, making it practically unique among burr puzzles. Two such solutions appear in Fig. 84. The one on the left is the standard six-piece burr, using five

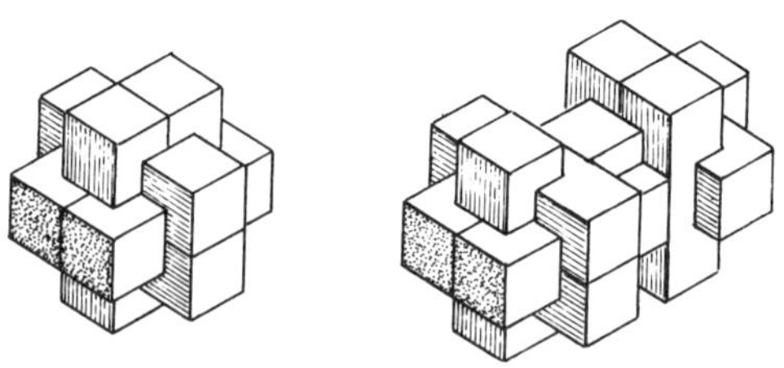

Fig. 84

regular pieces and one key piece. Note the unusual symmetry of the one on the right, which uses nine regular pieces and one key piece. There is another neat solution, left for the reader to discover, that uses 16 regular pieces and has fourfold symmetry.

In order to be satisfactory, the pieces must be made accurately, with the length exactly three times the width. A version of this puzzle is produced by Pentangle under the name *Squirrel Cage*.

Chapter 7

The diagonal burr

All of the burr puzzles described in the previous chapters have been orthogonal, i.e. rectilinear, Cartesian, with right angles. They are the most familiar and the easiest to visualize, analyse, explain, and make (but not necessarily to solve!). The time has now come to venture beyond the comfortable world of right angles and explore the wondrous geometry of the diagonal burr.

The diagonal burr can be regarded as a standard six-piece burr in which all of the sticks have been rotated 45 degrees, with the notches V-shaped rather than square. The six pieces shown in Fig. 85 represent one version. The piece with no notches is of course the 'key' which slides in last to complete the assembly. Two of the other pieces have an extra notch to accommodate it. The reader can probably solve the puzzle mentally by studying the drawings. It is also easy to whittle a rough model from square sticks of some soft wood.

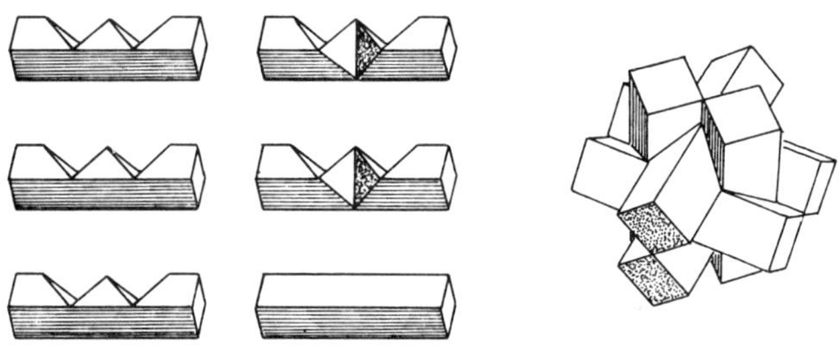

Fig. 85

After having solved this puzzle one way or the other, now make the surprising discovery that the 'key' is not really a key piece at all, but more properly called a pseudo-key. It need not go in last or come out first. In fact, it need not even be used. The burr can be assembled in total symmetry using six identical pieces with two notches each by mating two mirror-image halves of three pieces each (Fig. 86).

Like its orthogonal counterpart, the origin of the diagonal burr is a mystery. The earliest US Patent is No. 393 816 to Chandler in

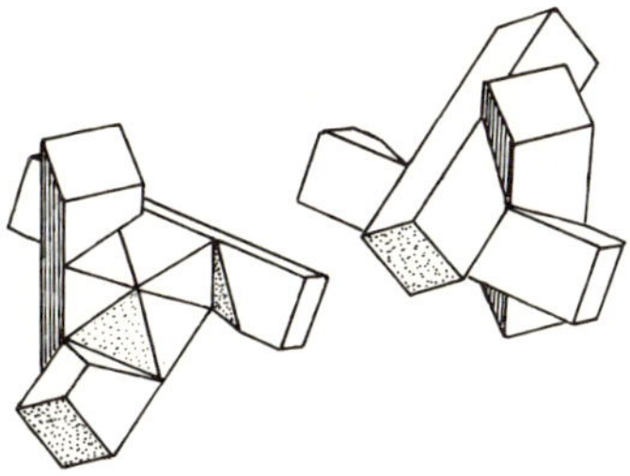

Fig. 86

1888, but it shows a more complicated version with sliding key. The earliest record of the symmetrical version appears to be US Patent No. 779 121 to Ford in 1905. Curiously, in his patent description, Ford shows a very awkward method of assembly rather than the simple mating of two halves.

The diagonal burr has also had its share of variations over the years. As you might expect, a favourite theme of puzzle inventors has been to increase the number of pieces, which is quite easy to do with this type of arrangement. Carrying this to the extreme, US Patent No. 774 197 to Pinnell in 1904 (Fig. 87) shows a horrendously complicated assembly of 102 diagonally notched sticks. (The patent notes that no model was submitted!) Another variation has been to enclose the burr in a spherical outer shell (US Patent No. 766 444 to Hoy in 1904 and US Patent No. 1 546 025 to Reichenbach in 1925).

Someone, somewhere, perhaps in the mid-19th century, made the marvellous discovery that the ends of the diagonal burr sticks can be bevelled to produce a puzzle that, when assembled, is the first stellation of the rhombic dodecahedron. According to puzzle collector and historian Jerry Slocum, a puzzle of this sort was sold as early as 1875. The only patent on it that the author is aware of is Swiss Patent No. 245 402 to Iffland in 1946.

The word *intriguing* is used frequently throughout this book to describe various polyhedral dissections, but none can outshine the brilliance of this simple dissection (Fig. 88). From one point of view, it may be regarded as a diagonal burr puzzle in which bevelling the ends of the pieces produces a totally unexpected and beautiful new shape. From another point of view, it is a surprising dissection of the stellated rhombic dodecahedron into six identical pieces that amazingly assemble and interlock! It has more interesting properties, too: when viewed along one of its fourfold axes of

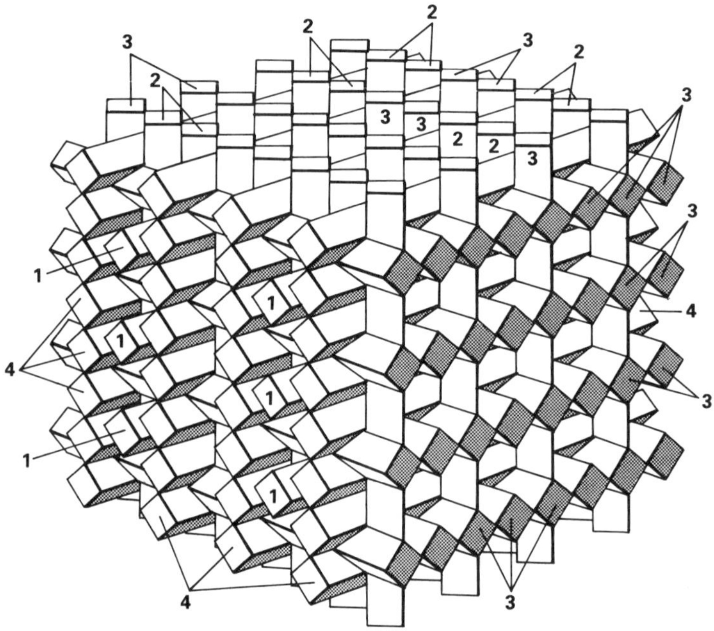

Fig. 87

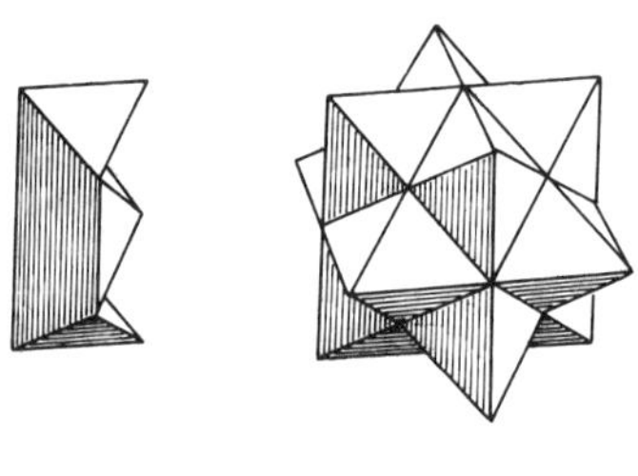

Fig. 88

symmetry it is square, while along one of its threefold axes it is the Star of David. And perhaps most surprising of all, it is a space-filling solid.

Although a rough model of the diagonal burr is easily whittled from soft wood or sawn by hand, to be entirely satisfactory it should be made very accurately, for which power tools and jigs are required. It is also quite susceptible to changes in humidity when made in wood, so stable woods should be used. It is often manufactured in plastic, which overcomes this problem.

The solution to the diagonal burr is so easy as to be barely a puzzle, but the stellated version does require some dexterity and patience, especially when accurately made with a tight fit. A problem with the stellated version in wood is that the sharp ends of the pieces are across the grain and easily broken. This can be corrected by making each piece of three blocks glued together. Although woodworking techniques are discussed later in this book, a method for making these blocks will be explained briefly here as an aid to understanding their geometry. This will be easiest if the reader can actually saw some out, but perhaps others can imagine doing it.

As shown in the drawing in Fig. 89, each puzzle piece consists of a six-sided centre block to which are attached a pair of tetrahedral end blocks. The six-sided centre blocks are easily made as follows: start with uniform square sticks of any convenient size—say one-inch square. Make a V-shaped cradle (Fig. 90) that holds the sticks at a 45-degree angle of rotation and slides in the mitre grooves of a table saw at an angle of 45 degrees when viewed from above. Make a diagonal cut on the end of the stick. Then rotate the stick 180 degrees and advance it in the cradle exactly the correct amount to produce a pyramidal point in the centre when the second saw cut is made. Continue in this manner to make additional blocks without waste. This very useful building block will be used frequently in the chapters that follow and will be referred to as the *six-sided centre block*.

Fig. 89

The two end blocks are made using the same square stock in the same cradle, except that the stock is advanced a shorter distance when making the second cut, and there will be a piece of waste for each one made. These are referred to as *tetrahedral blocks*. They are not regular tetrahedra—two opposite dihedral angles are right angles and the others are 60 degrees. These are also useful building blocks, both practically and mathematically. Many of the geometrical shapes to be discussed in the next three chapters can be regarded as made up of these blocks. For example, the six-sided centre block contains six of these units, the whole piece therefore

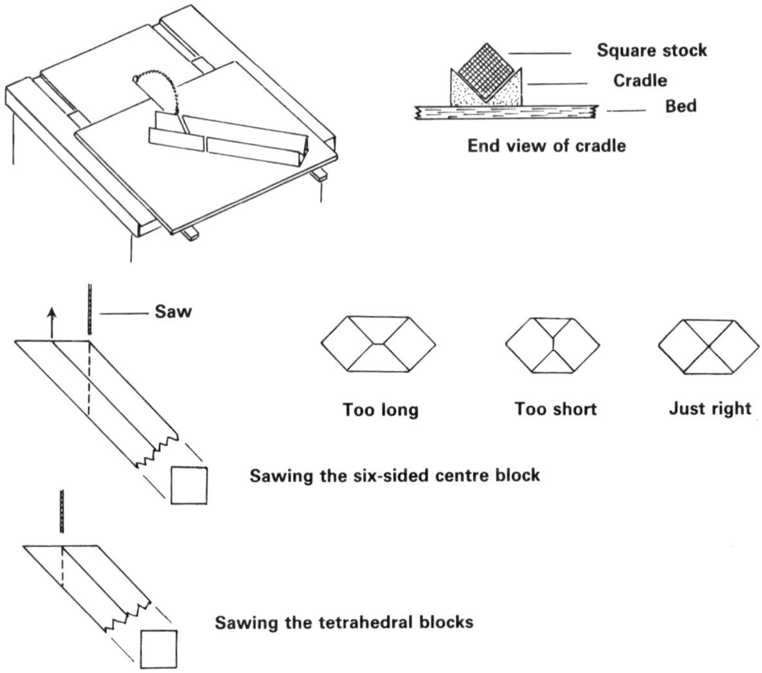

Fig. 90

eight, and the entire puzzle 48. The rhombic dodecahedron is made up of 24 of them, with each stellation containing two additional units.

Chapter 8

The rhombic dodecahedron and its stellations

The number of ways that sticks can be arranged symmetrically in space is very limited. It is convenient to examine this question in terms of unnotched straight sticks. The standard six-piece burr can be regarded as a cluster of six rectangular sticks to which parts have been added (or removed) to achieve interlock and other interesting features. The Pin-Hole Puzzle is an even better example. The hollow space in the centre is cubic. In any symmetrical arrangement of straight sticks totally enclosing a hollow centre, a little thought or experimentation will show that the faces of the enclosed hollow centre must be rhombic (or square). There are only three isometrically symmetrical solids with such faces—the cube, the rhombic dodecahedron, and the triacontahedron (Fig. 91).

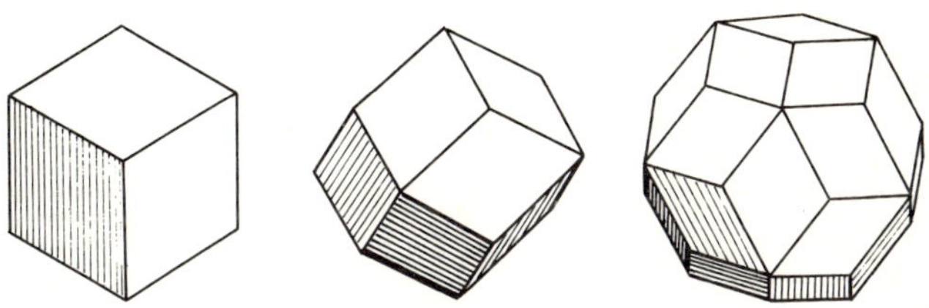

Fig. 91

The rhombic dodecahedron has 12 identical rhombic faces. It can be visualized as the solid that results when the edges of a cube are sufficiently bevelled at 45 degrees (Fig. 92a). It is one of very few symmetrical solids that pack to fill space, two others being the cube and truncated octahedron. Like the cube, it has three fourfold axes of symmetry, four threefold axes, and six twofold axes (Fig. 92b). When viewed along any of its fourfold axes it appears square in profile, while along any of its threefold axes it appears hexagonal (same as the cube).

The rhombic dodecahedron can be totally enclosed by a

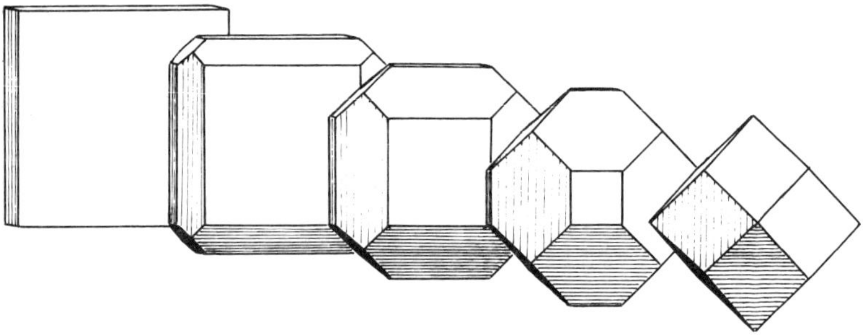

Fig. 92a

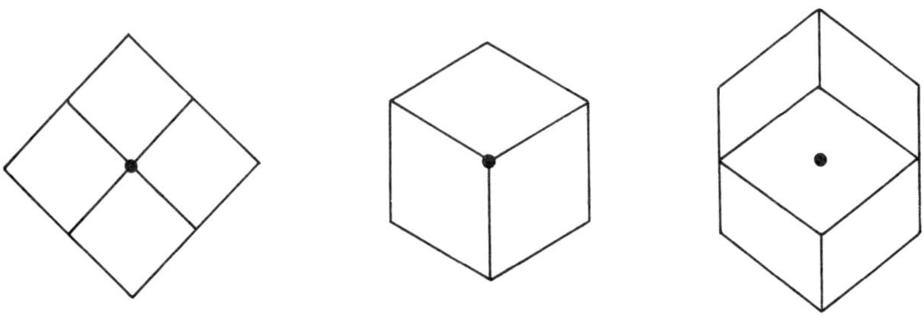

Fig. 92b

symmetrical cluster of 12 sticks having equilateral-triangular cross-section, a property not only intriguing but of great practical significance. This arrangement has a pair of mirror-image forms, as shown in Fig. 93.

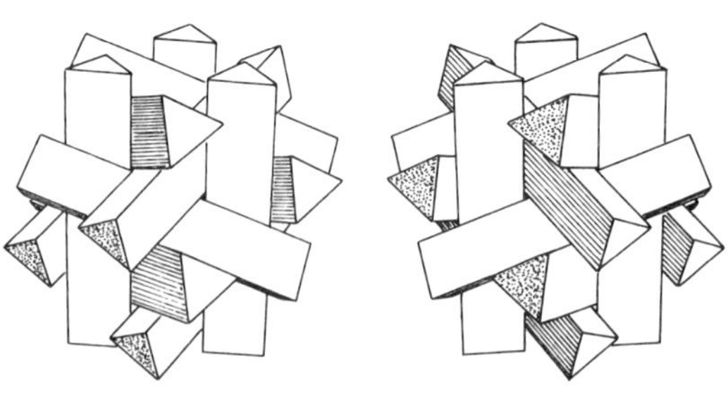

Fig. 93

Theory of interlock

At first glance, the nest of sticks shown in Fig. 93 may appear to be self-supporting. Any attempt to assemble it without tape or rubber bands will immediately dispel this notion as the sticks tumble into a heap. A useful tool for the puzzle designer would be some way of analysing such geometrical arrangements to determine if they are interlocking or not, or even possible to assemble. If the arrangement is totally symmetrical, there is a simple way to do this, as follows:

To take a trivially simple example to start with, consider the Pin-Hole Puzzle without the pins and holes. In its assembled condition as shown (Fig. 94), the two ends of each piece rest flat against two others, and the ends of yet two others rest flat against it. Now move each piece by some incremental distance directly away from the centre, and note that they become separated from each other. This is sufficient to show that the structure is non-interlocking and will easily fall apart.

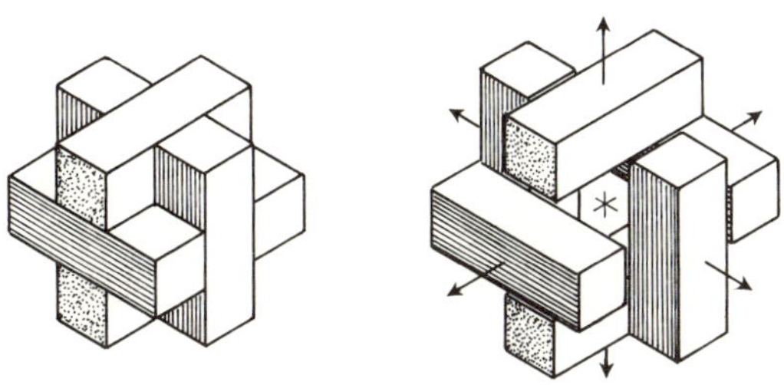

Fig. 94

To take one more trivial example, consider a standard six-piece burr (Fig. 95) made up of six identical pieces like notchable piece no. 2 (page 59). Applying this same test, we see that there is interference between the parts and therefore the burr is impossible to assemble.

Now for a more practical example, consider the diagonal six-piece burr (Fig. 96). As each piece is moved an incremental distance away from the centre, there is neither interference nor separation as the mating faces slide parallel to each other. Therefore it is an assemblable interlocking configuration.

This useful theory of interlock can be applied most easily by using elementary vector analysis. If the radial movement of one piece is represented by vector A and that of a neighbouring piece

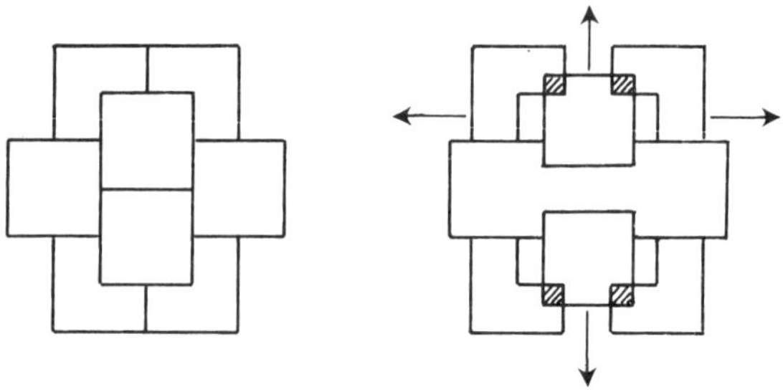

Fig. 95

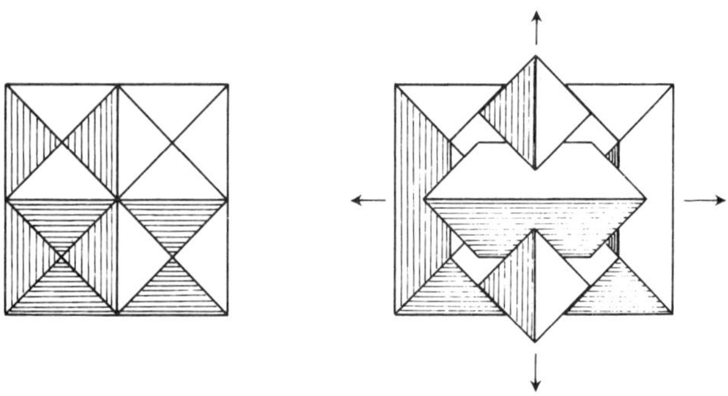

Fig. 96

by vector B, then the relative motion of the two is vector A–B, and any sliding surface in an assemblable interlocking puzzle must be parallel to it. Or make a scale drawing of the puzzle and use methods of descriptive geometry. With a little practice and good spatial perception or a model to work with, most of the assemblies discussed in this book are easy to analyse. Applying this theory to the nest of 12 triangular sticks, it is easy to show that they are non-interlocking.

How might the 12 triangular sticks be made into an interlocking assembly? One way would be to use notched sticks, as in the burr puzzles. That scheme will be considered in chapter 13. Another way is as follows: instead of leaving the centre hollow, imagine it filled solid with a rhombic dodecahedron. Now dissect that rhombic dodecahedron into six identical blocks having the shape of squat octahedra, and use each one of them as a centre block for joining the triangular sticks together in pairs, as shown in Fig. 97.

If the theory of interlock is applied to this new six-piece puzzle configuration, it is found to be an interlocking assembly. It can be

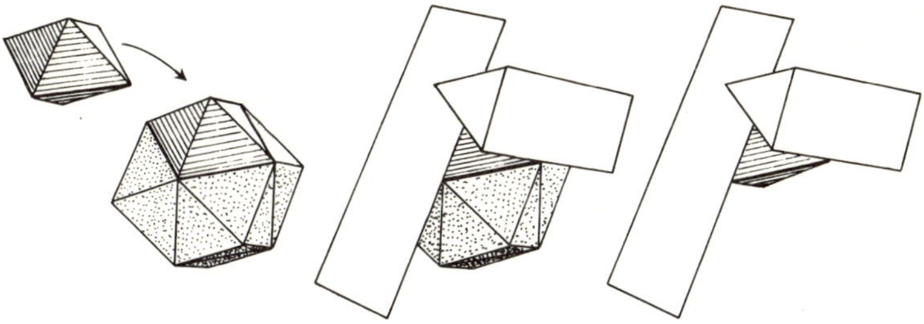

Fig. 97

slid apart along any one of its four sliding axes, independently or concurrently. Unlike the diagonal burr, it separates into two halves that are quite dissimilar, even though each half is composed of three identical pieces and the completed assembly is symmetrical.

Stellations

If both ends of all 12 sticks are now cut off at the appropriate angle (Fig. 98), an amazing transformation occurs and the assembly becomes the third stellation or the rhombic dodecahedron. (It is assumed that the reader has some familiarity with polyhedra and stellations. If not, any mathematics library should have a book on the subject.) Remove the equivalent of two tetrahedral blocks more from both ends of the sticks, and lo—the *second* stellation of the rhombic dodecahedron appears. Many intriguing intermediate forms are also possible by removing the equivalent of only one tetrahedral block, or by removing them selectively from certain ends. The biggest surprise occurs when yet two more tetrahedral

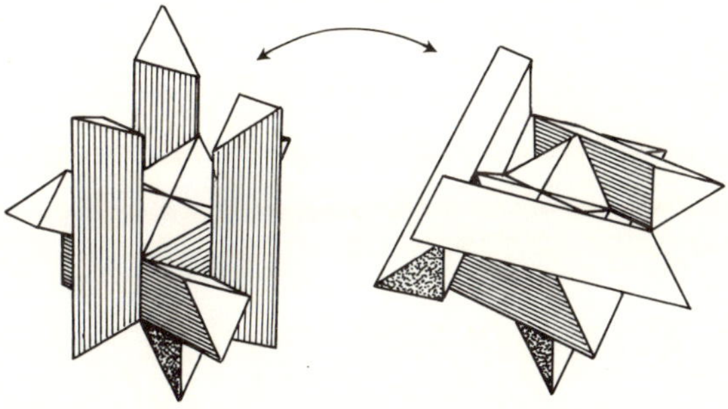

Fig. 98

units are removed from all the ends, producing the now familiar *first* stellation again (Fig. 99). Is this not amazing? It can be made not only from six square sticks with ends bevelled but also from 12 triangular sticks!

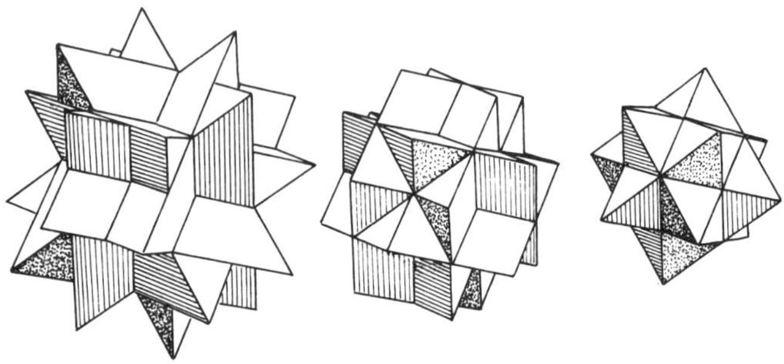

Fig. 99

The second stellation

Another surprise! Having now seen that the second stellation of the rhombic dodecahedron can be constructed by an interlocking assembly of 12 triangular sticks, would you believe that it too can be constructed (more easily, in fact) by an interlocking assembly of six pieces made from *square* sticks? Start with 18 six-sided centre blocks and join them in threes as shown in Fig. 100 to make six identical puzzle pieces, which assemble into the interesting interlocking polyhedral shape shown.

Now if V-shaped notches are made at both ends of each puzzle

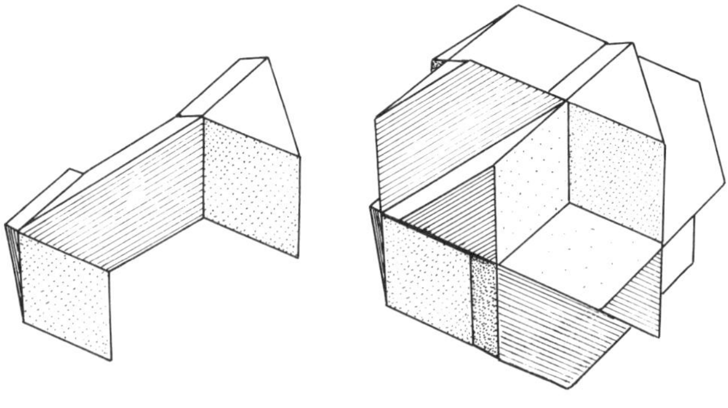

Fig. 100

piece in the model shown in Fig. 100, the second stellation is produced, as shown in Fig. 101. As a practical matter, rather than cut notches in the end blocks, it is easier to form them by gluing two suitable blocks together, both of which are easily made from square stock using the saw jig shown on page 78. These are very useful building blocks and will be used frequently in the next two chapters. One of them is a rhombic pyramid, and the other is a five-sided block having the shape of a skewed triangular prism, hereafter referred to as *prism block* for short.

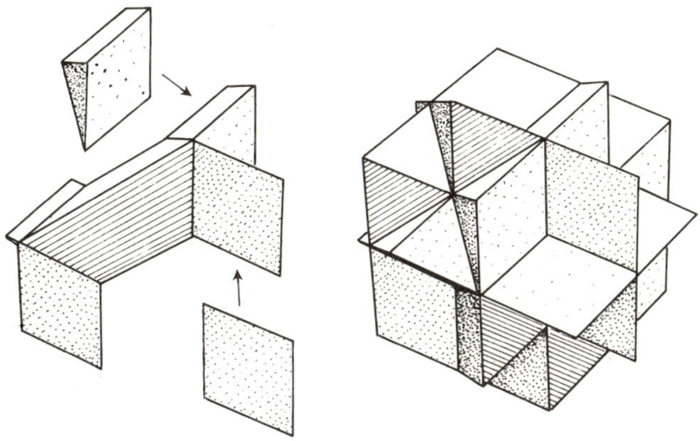

Fig. 101

All of the above models assemble by first forming two halves of three pieces each and then mating the two halves. Unlike those made with triangular sticks, these two halves are mirror images of each other.

The Four Corners Puzzle

In the second stellation model (Fig. 101), if the 12 rhombic pyramid blocks are omitted, the result is the simple but intriguing puzzle shown in Fig. 102. Its six identical pieces are assembled in the usual way of mating two halves, which in this case are dissimilar. The assembled shape is intermediate between the first and second stellation, and it has the symmetry of a tetrahedron. It serves as the skeleton for many other more complicated puzzles to follow. It will be referred to as the Four Corners Puzzle, the name by which a four-colour version of it was once produced.

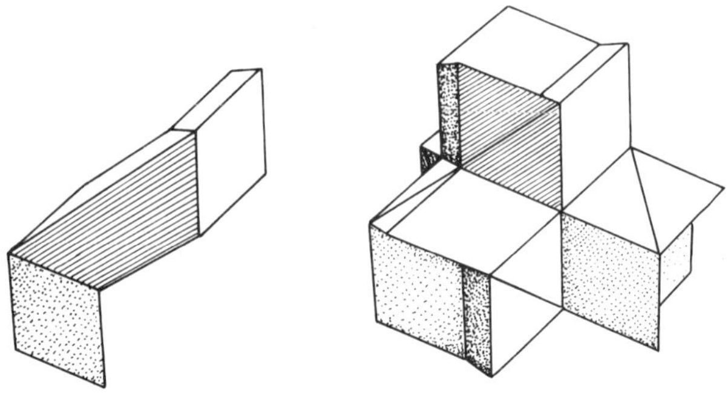

Fig. 102

Colour symmetry

The Four Corners Puzzle is a good example of an interlocking structure with an intriguing geometry and attractive shape but which is trivially simple as an assembly puzzle. To make it also challenging, the concept of colour symmetry is introduced. Imagine the end blocks coloured four different colours as indicated in Fig. 103.

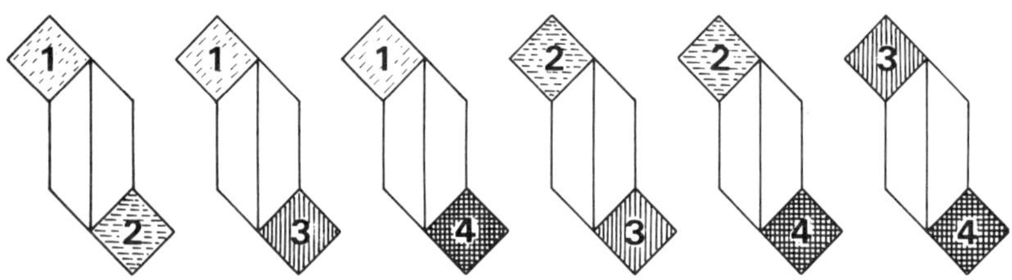

Fig. 103

Problem: assemble the above pieces in colour symmetry. Advanced problem: discover *all* the possible ways of assembling these pieces in colour symmetry. In order to solve this problem, we must first define exactly what is meant here by *colour symmetry.* When a multicoloured polyhedral puzzle is said to be assembled in colour symmetry, it meets the following test: choose any colour and change it to black. Change all the other colours to white. No matter which colour was changed to black, the result is the same, and the black pattern has an axis of symmetry.

The four different ways in which the Four Corners Puzzle can be assembled in colour symmetry are represented in Fig. 104 in black

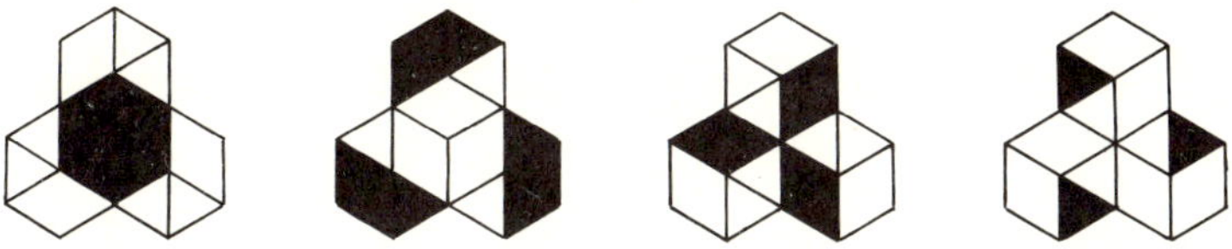

Fig. 104

and white. The one on the left, in which each 'corner' is a solid colour, is the easiest and most obvious and is how the puzzle got its name. Each has a pair of solutions.

Finally, to extract one more bit of recreation from this puzzle, discover the 24 ways of assembling it such that the patterns of all four colours are identical but *not* symmetrical. You may skip the 3808 ways that do *not* have either property. Hint: in general, these colour symmetry problems are not the type that one solves by trial and error. One must try to discover the principles involved and the simple rules that transform one solution into another. You may not even need the physical pieces.

The second stellation in four colours

Continuing in the same vein, shown in Fig. 105 are the six puzzle pieces for a four-colour version of the second stellation. Can you discover the logic of this colouring scheme? Hint: compare with the previous puzzle. Also note that the colouring produces reflexive pairs, as indicated by the broken line.

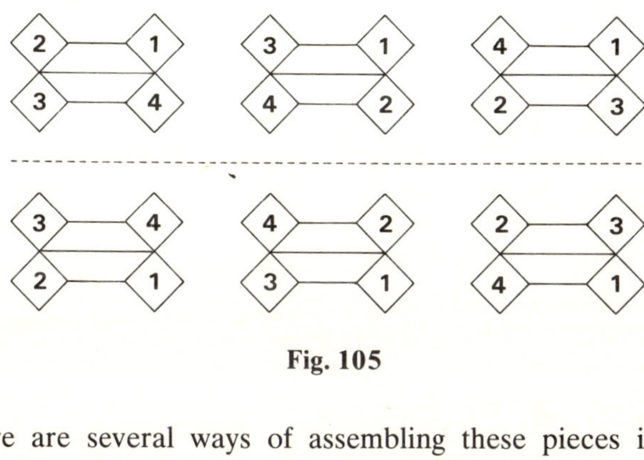

Fig. 105

There are several ways of assembling these pieces in colour symmetry, as shown in Fig. 106. The simplest and most obvious is with each of the eight hexagonal dimples a solid colour, with like

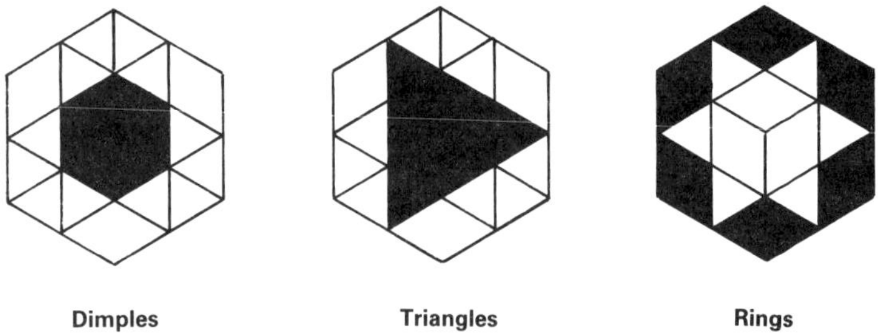

Fig. 106

colours opposite. Another way is with four triangles of solid colour. The most elegant is with four hexagonal rings of solid colour intersecting each other around the outside. The other ways are left for the curious reader to discover. Trying to solve all of these fascinating colour patterns and being able to switch from one to another can be quite confusing and entertaining.

The third stellation in four colours

There are many different ways of dissecting the various stellations of the rhombic dodecahedron into six identical interlocking pieces, and no purpose would be served by listing them all. Just one more example will be mentioned in this chapter—a simple dissection of the third stellation that lends itself beautifully to a multicolour puzzle.

The construction of each puzzle piece from a six-sided centre block and four triangular stick segments is illustrated in Fig. 107. When assembled, the puzzle has the appearance of twelve triangular sticks, even though each stick is broken in two, with the two halves belonging to two different puzzle pieces. The pieces are coloured as shown. The problem is to assemble the puzzle such that each apparent group of three parallel sticks is one colour (see cover). There are four solutions.

These are but a few of the many interesting multicolour problems that are possible with puzzles of this sort. For example, the three described above all use four colours. Other possibilities exist using 2, 3, 6, 8, or 12 colours. The woods can be stained or painted different colours, but some of the most beautiful effects are obtained by using brightly coloured exotic woods in their natural state. More multicolour puzzles will be described in later chapters.

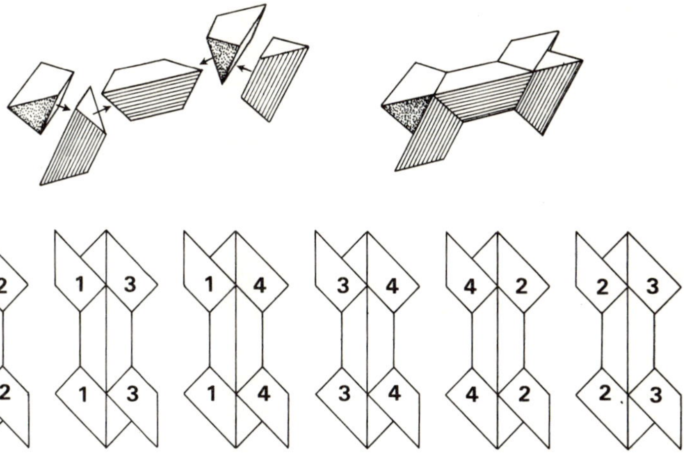

Fig. 107

Chapter 9

Polyhedral puzzles with dissimilar pieces

In the previous chapter, simple totally symmetrical dissections were transformed into puzzling problems through the use of colouring. We will now explore the combinatorial complexities created by making the pieces dissimilar in shape.

The permutated second stellation

As already explained on page 85, the second stellation shape is obtained by adding twelve rhombic pyramid blocks to the Four Corners Puzzle. As each block is added, there is a choice of two surfaces to which it can be attached. The six puzzle pieces shown in Fig. 108 represent all of the non-symmetrical ways of attaching such blocks. The added blocks are shown shaded. By an amazing coincidence, this just happens to use the required twelve blocks and yield the required six puzzle pieces. This quite satisfactory combinatorial puzzle has two solutions of moderate difficulty.

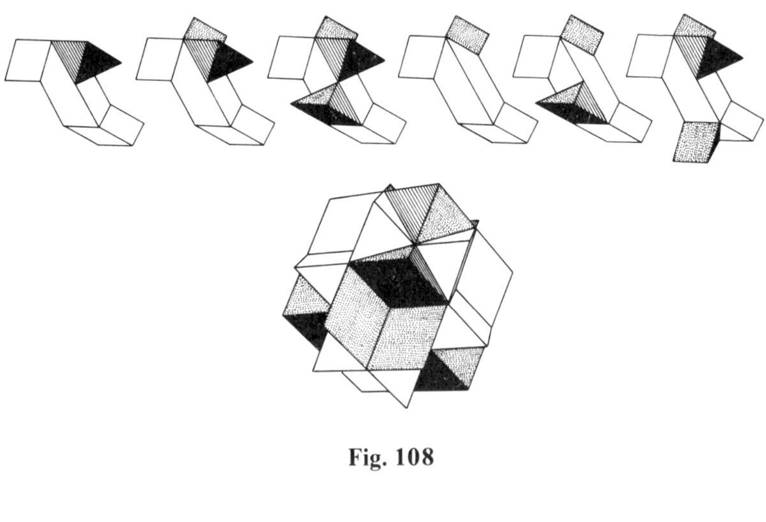

Fig. 108

The permutated third stellation

The fortuitous geometrical circumstance described above, which allows the second stellation to be assembled from a fully permutated set of puzzle pieces, can be exploited in many other polyhedral shapes. The analogous set of pieces for the third stellation is shown in Fig. 109. It likewise has two solutions and four sliding axes. Theoretically, it should be no more difficult than the second stellation, but it is probably slightly more confusing because of the greater irregularity of the pieces.

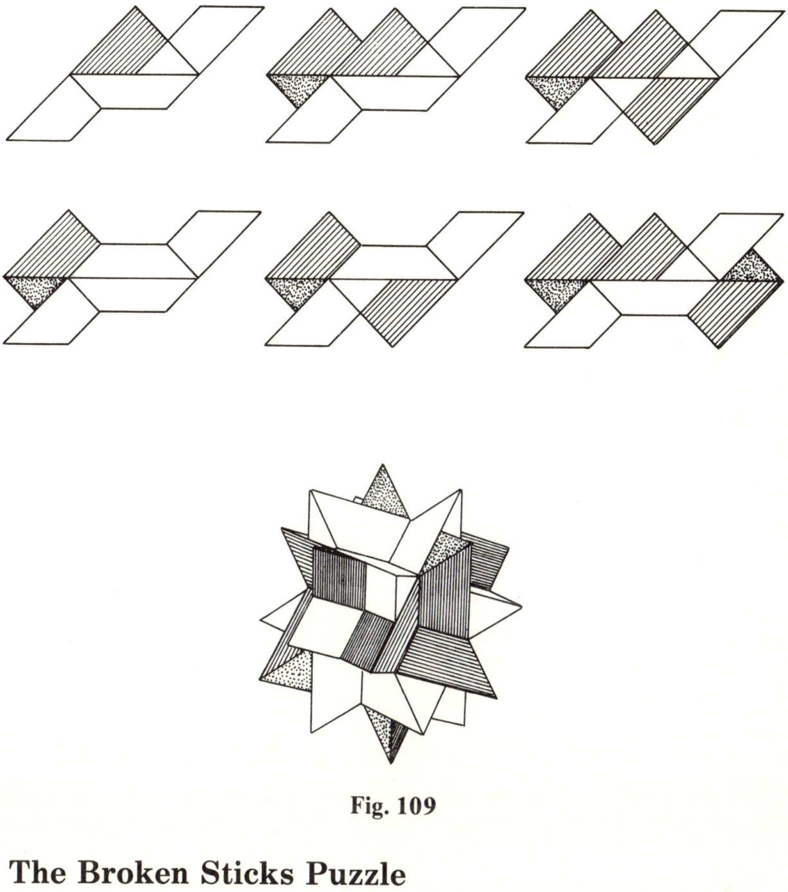

Fig. 109

The Broken Sticks Puzzle

Carrying the development of the two puzzles described above one step further, by lengthening the triangular stick segments even more, one arrives back at an assembly that resembles the nest of twelve triangular sticks, except that now some of them are broken in two internally. The six dissimilar puzzle pieces are shown in Fig. 110. Because of the extra length of the arms, they now interfere with each other during assembly. By an extraordinary coincidence, this results in one of the two solutions being impossible to assemble and three of the four sliding axes being blocked. Thus the puzzle has only one solution and one possible sliding axis of assembly. Consequently it is very difficult.

This remarkable puzzle design has been described as though it were a coincidence of four coincidences. But is it really? The term *coincidence* would seem to imply chance or luck, whereas this is simply a mathematical reality of the way things are in this world. The only luck involved was that of the person discovering it. Or was it? Perhaps the universe itself is the ultimate example of an improbable coincidence. Why are things as they are?

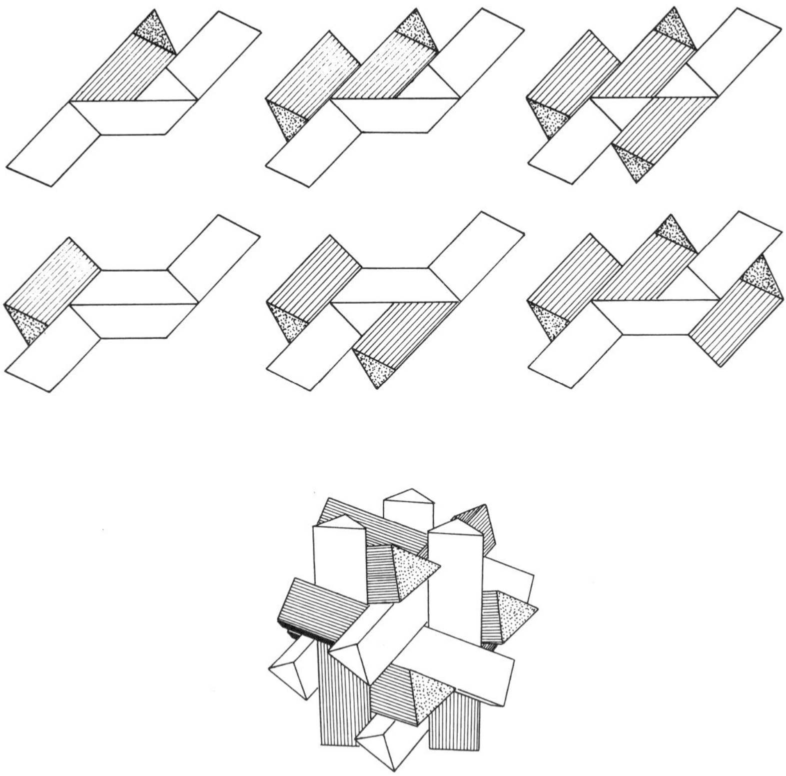

Fig. 110

The augmented second stellation

With the addition of yet 12 more rhombic pyramid blocks to the permutated second stellation of the rhombic dodecahedron, it is transformed into the intriguing polyhedron shown in Fig. 111, which is an intermediate form between the second and third stellations. More importantly, these added blocks have the same practical effect as did the lengthening of the arms in the Broken Sticks Puzzle. That is, one of the two solutions is eliminated and three of the four sliding axes are blocked. Besides forming an attractive polyhedral sculpture with simple clean lines, especially when made of contrasting exotic woods, its snugly interfitting pieces do not require much dexterity to assemble. All things considered, this is a most satisfactory puzzle.

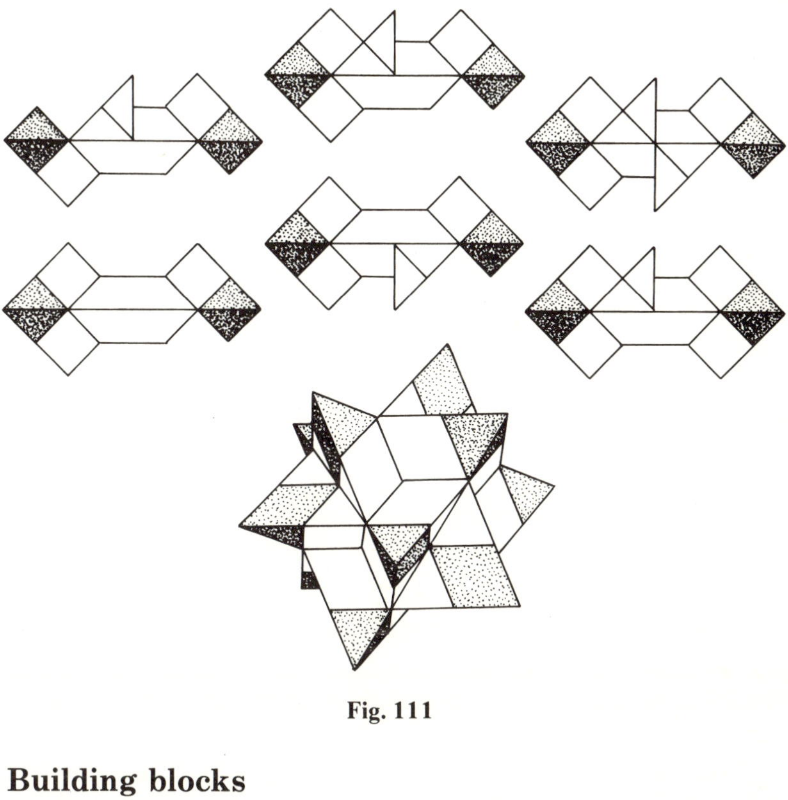

Fig. 111

Building blocks

In this and the previous two chapters, various polyhedral blocks derived from dissections of the rhombic dodecahedron have been employed for building up puzzle pieces. If the geometry of these pieces is not entirely clear to the reader from the drawings alone, some hands-on experience with the blocks should help to clarify things. If the requirement for accuracy is set aside for the moment, they are all easy to make even with hand tools. This is a good point at which to summarize them.

For our purposes, the tetrahedral block is taken as the most basic unit, although of course it could be further subdivided *ad infinitum*. Since many of the blocks are made equally well from either square or triangular stock, this is conveyed with two sets of drawings in Fig. 112.

For further clarification, Fig. 113 shows rhombic dodecahedra dissected into these various shapes. The rhombic dodecahedron itself will be used as a basic building block in chapter 18.

With only these six building blocks, the number of simple ways in which they can be combined into interlocking puzzles is phenomenal. Just a few more examples will be shown in this chapter.

The Augmented Four Corners Puzzle

The single puzzle piece shown in Fig. 114 consists of a six-sided centre block to which have been attached two right-handed prism blocks. Six such identical pieces assemble into an interlocking configuration as shown, with three gaps in each of the four corners. These gaps are filled with 12 rhombic pyramid blocks. Each block can be attached to either of two adjacent pieces. There is one and only one way of attaching them whereby six dissimilar non-symmetrical puzzle pieces are created, as shown. They assemble one way only, with only one sliding axis, to make a very satisfactory puzzle. The appearance of the puzzle is enhanced by using a contrasting wood for the added blocks.

The Diagonal Cube Puzzle

Often, two polyhedral dissections may be internally similar and functionally identical as assembly puzzles, even though their external appearances are quite dissimilar. Here is a good example.

The single puzzle piece shown in Fig. 115 consists of three six-sided centre blocks joined together. Six such identical pieces assemble into an interlocking configuration as shown, with three gaps in each of the eight corners. These gaps are filled with 24

T Tetrahedral block
Basic unit.
Made from triangular stock without waste or square stock with waste.

P Rhombic pyramid block
Two tetrahedral units.
Made from triangular stock without waste or square stock with waste.

R Right-handed prism block
Three tetrahedral units.
Made from either triangular or square stock without waste.

L Left-handed prism block
Three tetrahedral units.
Made from either triangular or square stock without waste.

O Squat octahedron block
Four tetrahedral units.
Made from square stock with waste.
Also made of two rhombic pyramid blocks.

C Six-sided centre block
Six tetrahedral units.
Made from square stock without waste.
Also made of two prism blocks.

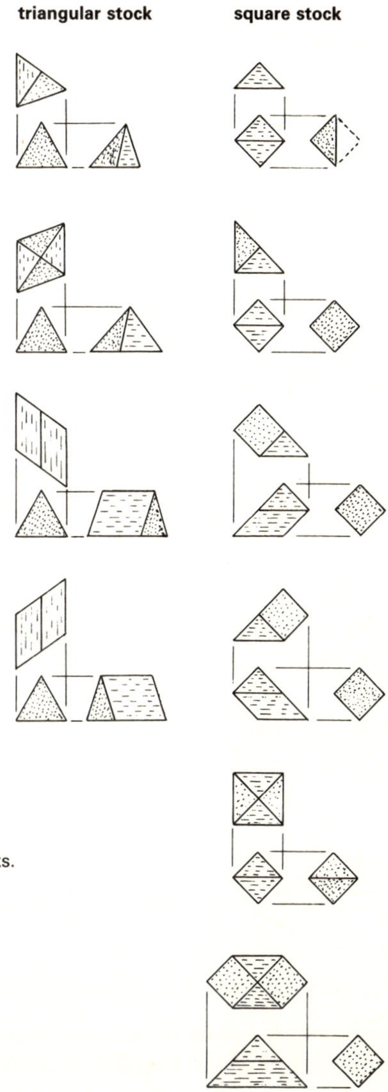

Fig. 112

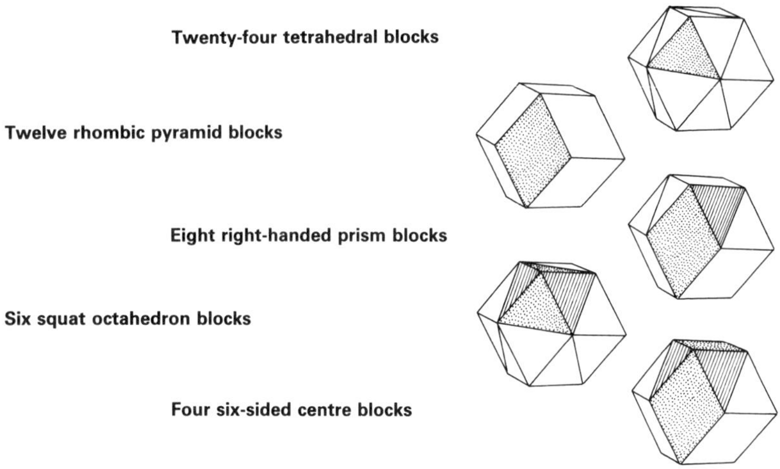

Fig. 113

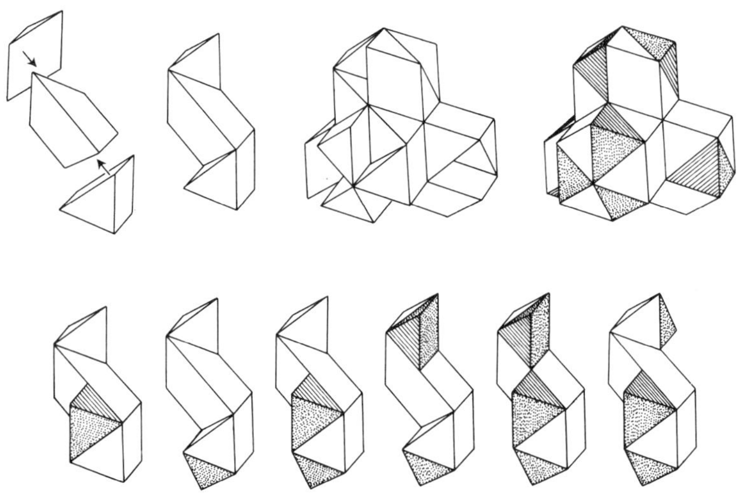

Fig. 114

rhombic pyramid blocks attached in such a fashion that six dissimilar non-symmetrical puzzle pieces are created. The six sides of the assembled puzzle are then truncated into square faces, making it cubic. The sliding axis for the first step of disassembly is an internal diagonal of the cube and can be rather tricky to locate. Thus, this puzzle is entertaining both to solve and to disassemble. It

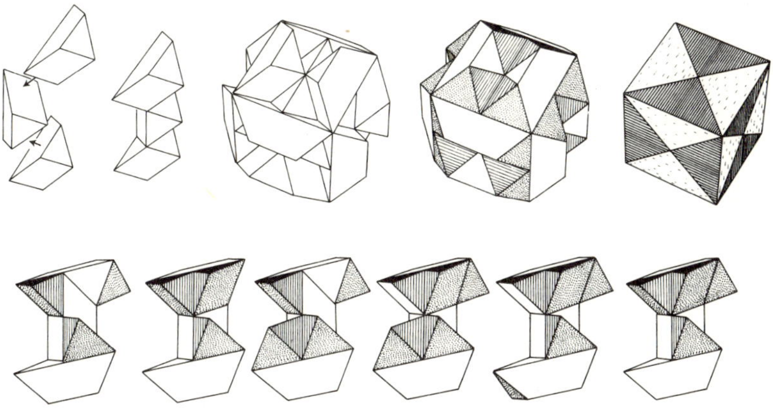

Fig. 115

is also quite attractive in two contrasting woods, with the six faces sanded and polished.

The Reluctant Cluster Puzzle

If the Diagonal Cube Puzzle is incidentally tricky to disassemble, this theme is carried one step further in the next example, which is made deliberately so.

Each puzzle piece consists of a six-sided centre block, to each end of which is attached another six-sided centre block and a prism block. In the version shown in Fig. 116, a right-handed prism block is at one end and a left-handed one at the other end, so that each piece has reflexive symmetry and all six pieces are identical. (Other

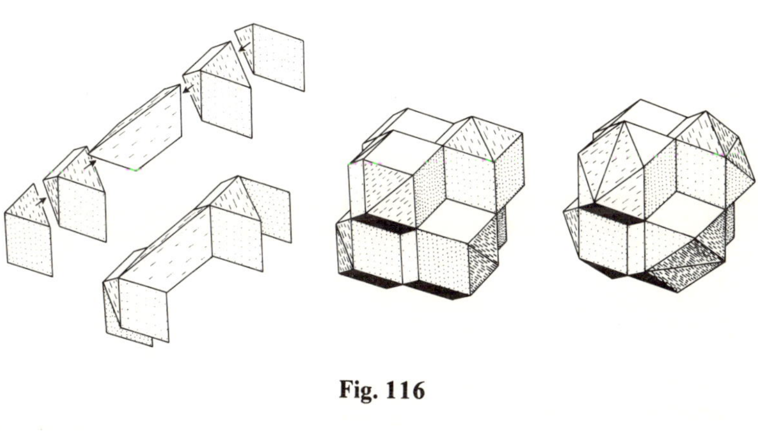

Fig. 116

more complicated versions are possible having dissimilar pieces.) The puzzle is assembled with no great difficulty by mating two mirror-image halves to form an interesting polyhedral shape having the appearance of a cluster of six rhombic dodecahedra. To *dis*-assemble it, however, three fingers of each hand must be placed in exactly the correct places and pushed in opposition to start the two halves apart along their one sliding axis. Even when made slightly loose so that it rattles in one's hands, random poking and prying will seldom dislodge the two halves. A truncated version is also shown.

Chapter 10

Intersecting prisms

The Four Corners Puzzle, especially the augmented version, has rather the appearance of four mutually intersecting prisms. By use of triangular stick segments in the construction, this effect is accentuated to create some interesting sculptural shapes that are also enjoyable interlocking puzzles.

The Hexagonal Prism Puzzle

The first example shown in Fig. 117 is directly analogous to the Augmented Four Corners Puzzle and has the appearance of four mutually intersecting hexagonal prisms. The six dissimilar puzzle pieces assemble one way only, with but one sliding axis. The triangular pattern in the hexagonal faces can be accentuated by using a contrasting wood for the permutated blocks.

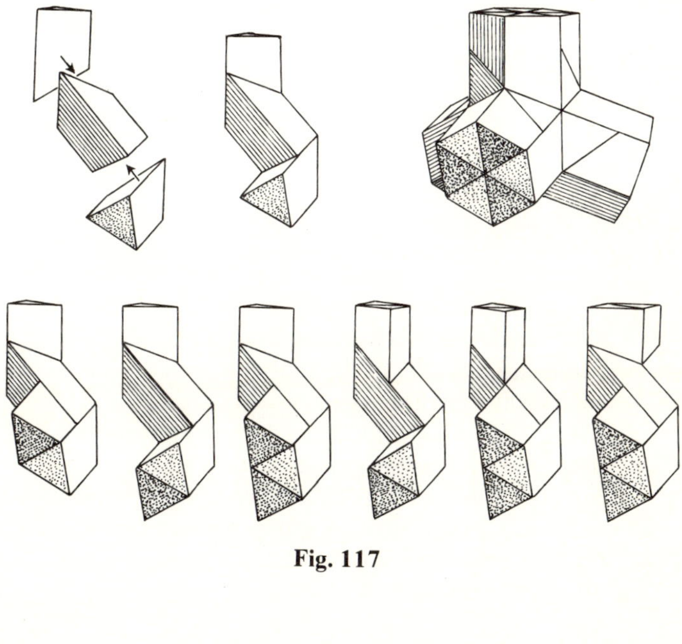

Fig. 117

The Triangular Prism Puzzle

By adding 12 more triangular stick segments to the Hexagonal Prism Puzzle, it is transformed into a most fascinating geometrical solid having the appearance of four mutually intersecting triangular prisms. With 42 blocks used in the construction, many design variations are possible depending upon how some of them are attached. One version having six dissimilar pieces is shown in Fig. 118.

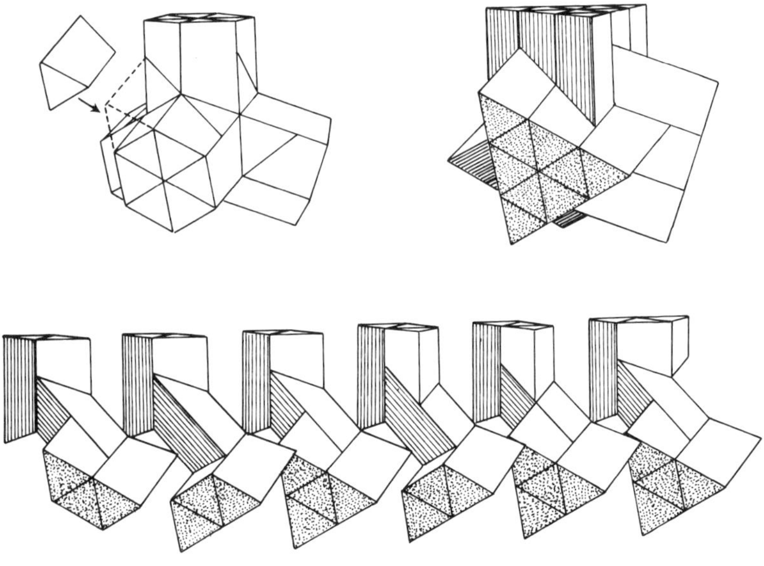

Fig. 118

The Star Prism Puzzle

By adding yet 12 more triangular stick segments to the Triangular Prism Puzzle, the prism faces assume the shape of the six-pointed Star of David. With 54 blocks used in the construction, a great many variations are possible in the individual pieces, all having the assembled shape shown in Fig. 119.

By adding even more blocks, other sculptural effects are possible, such as extending the prisms in the opposite directions as in the models illustrated in Fig. 120. Since these complicated constructions are more artistic than logical or mathematical, details are not given here but rather left to the craftsman's imagination.

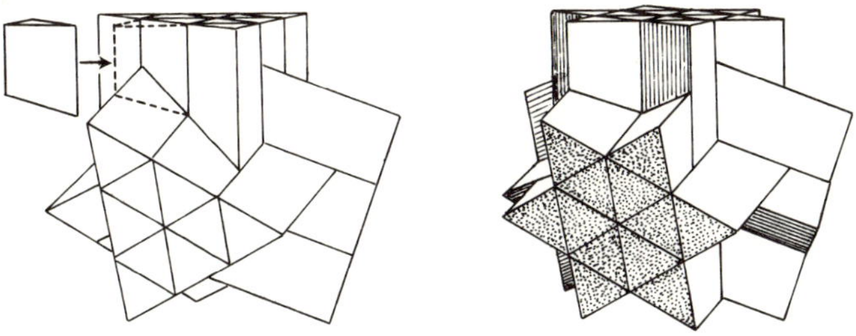

Fig. 119

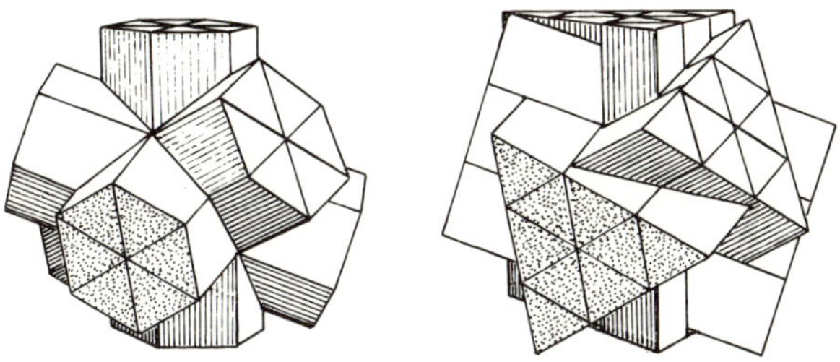

Fig. 120

Woodworking note

Many of the designs in this and the previous chapter consist of a basic skeleton of six identical interlocking parts to which are attached additional blocks, making the pieces dissimilar and non-symmetrical and the solution unique. These include the Broken Sticks Puzzle, Augmented Four Corners Puzzle, Diagonal Cube Puzzle, and all of the Prism family of puzzles above. The most satisfactory method for making any of these is to first make the six identical parts, assemble them tightly to form the skeleton, and then glue the permutated blocks in the appropriate slots. This assures a perfect fit every time the puzzle is assembled. The Triangular Prism Puzzle is made by gluing the 12 additional stick segments on to an assembled Hexagonal Prism Puzzle, and so on, as suggested by the illustrations. Use wax or waxed paper to prevent accidental glue joints. The end faces of the Prism Puzzles are sanded true in the assembled state. More woodworking information is given in chapter 23.

The Square Prism Puzzle

Getting back to basics, at the ancestral root of all these strange non-rectilinear dissections is the venerable diagonal burr. An interesting variation of the diagonal burr is one using sticks having isosceles-right-triangular rather than square cross-section, as shown in Fig. 121. The six identical pieces assemble into an intriguing shape having the appearance of three mutually intersecting square columns. When well-crafted of three contrasting woods, the effect can be quite pleasing. As an assembly puzzle, it is so simple as to be almost trivial. But humble parents sometimes have precocious offspring.

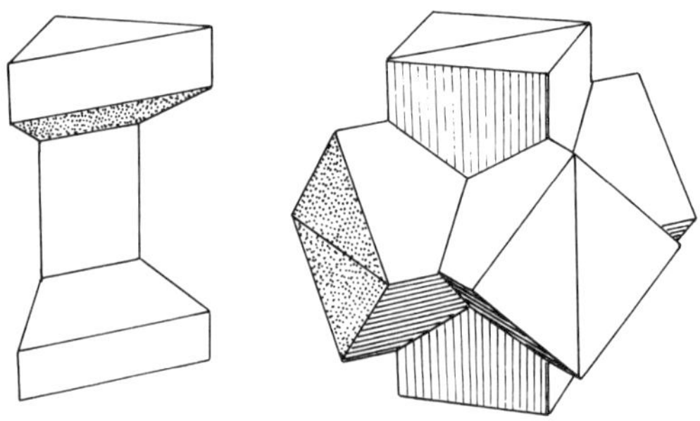

Fig. 121

The Three Pairs Puzzle

Now imagine each piece of the Square Prism Puzzle split in two longitudinally, resulting in 12 identical half-pieces. As an assembly puzzle, this additional dissection merely transforms it into more of a dexterity problem, which is certainly not a step forward. But now join these half-pieces in perpendicular pairs—three right-handed and three left-handed. Now assemble! This amazing puzzle, with its six simple pieces, has baffled experts. Even the name is a joke!

This puzzle was designed in 1973. Only about 200 have been produced, nearly all in mahogany. It is not difficult to make except that care is required to achieve an accurate fit, and stable woods must be used. The reader may wonder why it has not been mass-produced at low cost, perhaps of injection-moulded plastic, so that anyone may own one. That approach may not be such a good idea.

First of all, there is no way that plastic can compete with wood aesthetically. Without arguing this point and the reasons for it, anyone who has sold handicrafts for a living knows that it is so. In toy manufacturing, the bottom line is usually profit and quarterly earnings, so puzzles are usually made of cheap styrene, warped by shrinkage, tapered slightly for easy ejection, and cored out to further reduce costs. To recover the investment in the mould, hundreds of thousands must be made exactly alike, whereas the

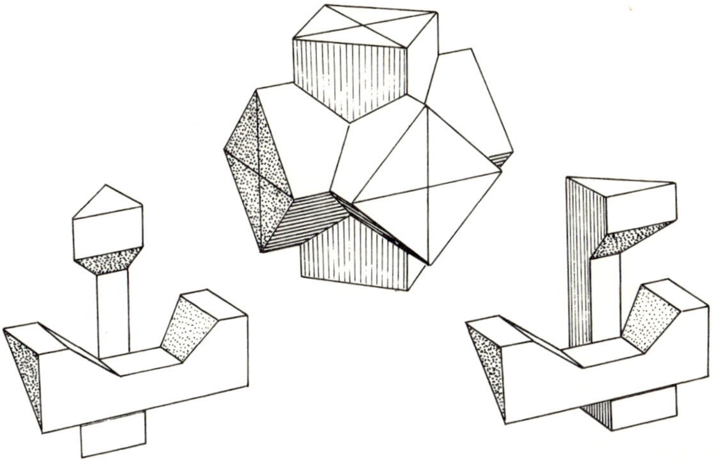

Fig. 122

designer-craftsman is always experimenting with variations and improvements. To reduce mould costs, compromises are made in design selection, especially avoiding those with all dissimilar pieces or requiring complicated moulds with side action. Also lost is the close rapport between designer and the public, when the manufacturer, jobber, and retailer all stand between. Perhaps some things, such as automobiles, are practical to manufacture only in large factories, but creative playthings crafted by hand are likely to bring more satisfaction to both maker and user.

Chapter 11

Puzzles that make different shapes

In chapter 1, it was shown that plane dissection puzzles have much greater recreational potential when the same set of puzzle pieces assembles into many different problem shapes. The only three-dimensional puzzles described thus far having this property have been a couple of the cubic block types that form different rectangular solids and the Squirrel Cage burr set. Dissections between two different common geometrical solids present horrendous difficulties and in most cases have been proven impossible. Yet the search goes on for three-dimensional puzzles that not only assemble into different geometrical shapes but also interlock. Two designs that partially succeed are given in this chapter.

The Star of David Puzzle

The six dissimilar and non-symmetrical pieces of the Star of David Puzzle are shown in Fig. 123. The 27 individual blocks required in their construction are all standard building blocks from the chart

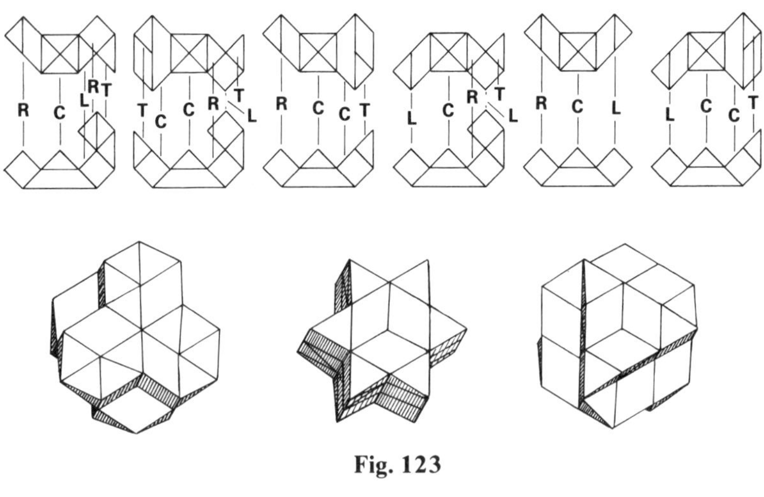

Fig. 123

on page 95 and are identified by letter. This unusual puzzle assembles into three different geometrical shapes having an axis of symmetry, as well as other nondescript shapes having no apparent symmetry. All solutions have just one sliding axis of assembly. In the solution from which the puzzle derives its name, the assembly axis does not coincide with the axis of symmetry. Consequently, one blindly tries various combinations looking for the solution. Even as the correct two halves are being mated to complete the assembly, it still looks like the sort of jumble one associates with 'abstract scupture'. But as they mesh together, suddenly there the solution is! This unusual and baffling puzzle presents a challenge for the skilled woodworker as well as the solver, for it is more difficult than most to fabricate well with a proper fit.

A puzzle in reverse

The Star of David Puzzle has a relative that is mechanically much simpler but just as entertaining. This puzzle was published by the author several years ago in the form of the three axially symmetrical solutions illustrated. All the reader was asked to do was to work backwards and figure out the design of the six identical puzzle

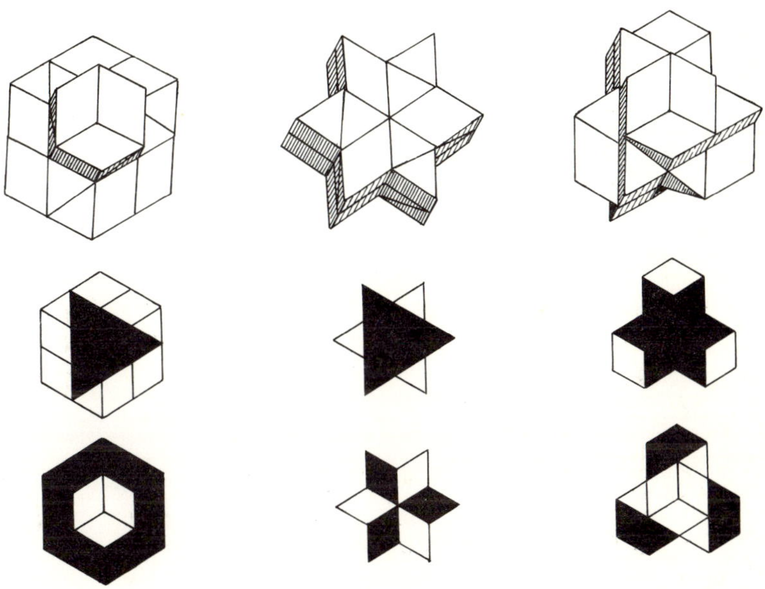

Fig. 124

pieces. Hint: each piece is made of three standard building blocks and has reflexive symmetry. But wait—why make it so easy? The pieces are bicoloured. Each of the three symmetrical solutions has two symmetrically coloured forms which are almost but not quite the inverse of each other, as shown in Fig. 124. *Now* what do the pieces look like?

Any reader who solves the above problem mentally or using pencil and paper should consider it quite an accomplishment. On the other hand, once you have the opportunity to play around with these polyhedral blocks in the flesh and experiment with fastening them together different ways, the solutions to this and other similar problems should all become perfectly obvious. Not only that, but in the process you will probably discover other interesting dissections as well.

Chapter 12

Coordinate–motion puzzles

Figure 125 shows a very simple amusement consisting of three identical rhombic blocks with pins and holes. Note that the only way it can be assembled is to start all three pins into the holes simultaneously and bring all three blocks together. This model is a simple example of a *coordinate–motion* puzzle. Such puzzles cannot be assembled sequentially, but rather at some stage of assembly they require the simultaneous manipulation of three or more pieces or groups of pieces. Unlike this simple example, such puzzles can be very baffling. One such puzzle has already been included surreptitiously in a previous chapter without being identified as such!

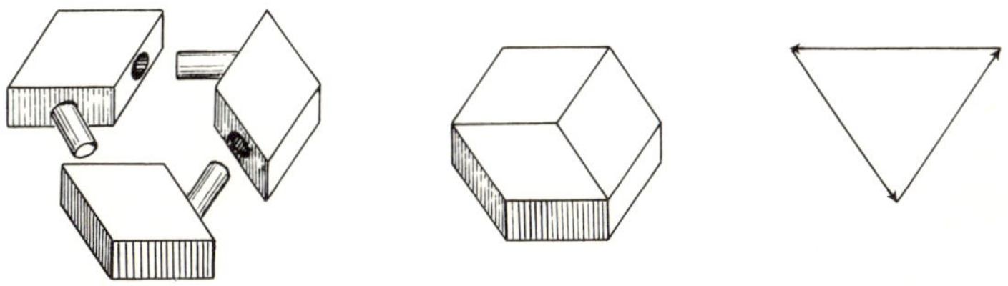

Fig. 125

The Expanding Box Puzzle

This simple novelty (Fig. 126) is a practical example of a coordinate–motion puzzle in three dimensions. Each of the six identical puzzle pieces is made up of a right-triangular prism centre block to which are attached a pair of rhomboid-prism end blocks. They assemble with no great mystery to form a hollow box, but if they are accurately made, some dexterity is required to get all six pieces aligned exactly right and mutually engaged. Once assembled, by holding on to opposite pieces, the puzzle can be made to expand almost to the point of collapse and then to shrink back together again. It is more of an amusement than a puzzle.

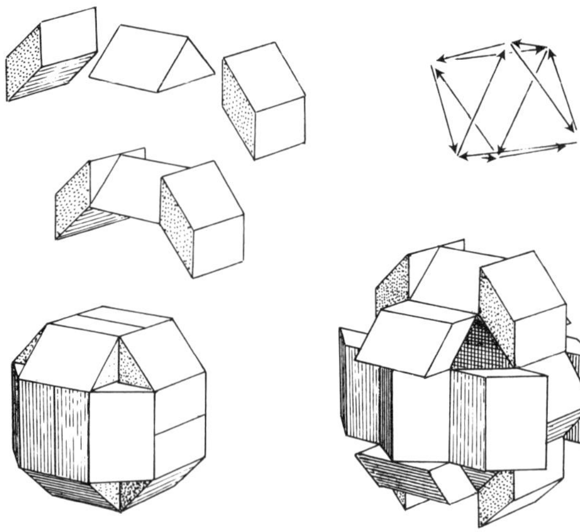

Fig. 126

The Rosebud Puzzle

The Rosebud Puzzle (Fig. 127) consists of six pieces, of which three are identical and the other three are their reflexive pairs, except that one piece has a small hole for a removable pin. This puzzle has two symmetrical solutions, one of which is decidedly more interesting than the other. Assembling it in the preferred solution requires the simultaneous manipulation of all six pieces in a manner that to begin with is not at all obvious and furthermore requires dexterity, patience, and some clever tricks.

A few of these puzzles were produced some years ago and sold unassembled. After sufficient time had elapsed and almost none had been solved, the customers were given the opportunity to purchase (for an outrageous price!) an assembly jig and directions. With these, it is easy. Without the jig, it can be done with patience, using tape and rubber bands. Without such aids, it has been done but borders on the impossible.

The reward for going to all this trouble is to discover the amazing mechanical action of the assembled puzzle. By placing three fingers of each hand in just the right places, it can be made to expand in a most fascinating manner, rather like the petals of a blooming flower. This effect is further enhanced by making the blocks of colourful woods with symmetrically matched grain. Then shift all six fingers slightly and the Rosebud closes again. By inserting the

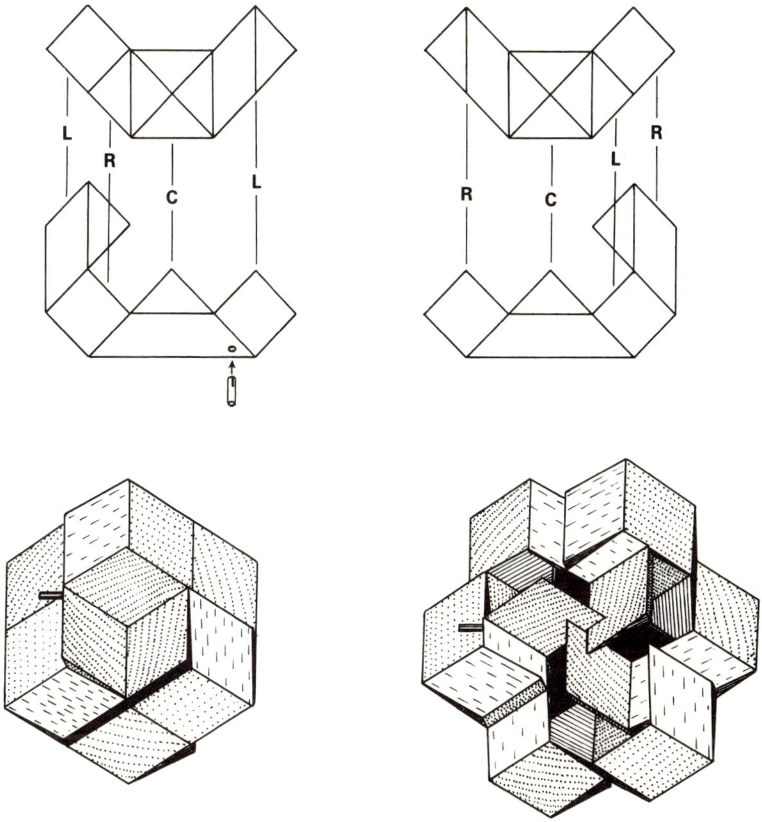

Fig. 127

pin into the hole after assembly, one can play with the puzzle in this manner with no worry that it will fly apart.

Standard building blocks are used in the construction of the puzzle pieces, as indicated. In order to be entirely satisfactory, they must be made very accurately with a smooth sliding action and must be strong.

The mechanical action of a coordinate–motion puzzle can be analysed and explained using a vector diagram. For instance, in the first example shown, if the three rhombic blocks are A, B, and C, their relative motions are vectors AB, BC, and CA, which must form a closed loop and add up to zero—in this case by an equilateral triangle. The vector diagram of the Expanding Box

Puzzle is an octahedron, and for the Rosebud Puzzle it is a triangular antiprism. The vector diagram for a coordinate-motion puzzle cannot be rectilinear. Why?

Although it is conceivable that one might design a coordinate-motion puzzle by studying vector diagrams, most such designs are just stumbled upon. The vector diagrams are presented here mostly as a curiosity.

Chapter 13

Puzzles using hexagonal or rhombic sticks

Refer back once again to the ubiquitous cluster of 12 triangular sticks on page 80. For the sake of variety, reduce them from triangular to hexagonal cross-section. They will still rest flat against each other. To hold them together, drill five holes in each stick and pin them together with 12 dowels (Fig. 128).

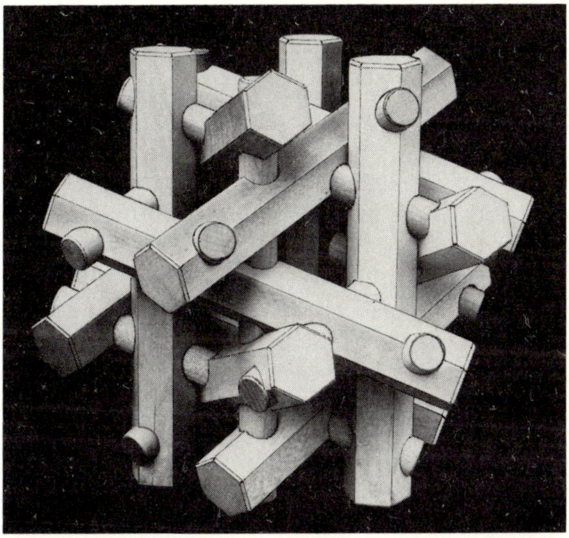

Fig. 128

Assembling this cluster of 12 hexagonal sticks and 12 dowels might be considered a puzzle of sorts—easy if an illustration is provided but perhaps not so easy otherwise. To make it into a more interesting puzzle, join some of the sticks and dowels to make elbow-shaped pieces (Fig. 129). The more elbows made, the harder the puzzle. With five elbows it is hard. With six elbows it is very hard. With seven, it is impossible.

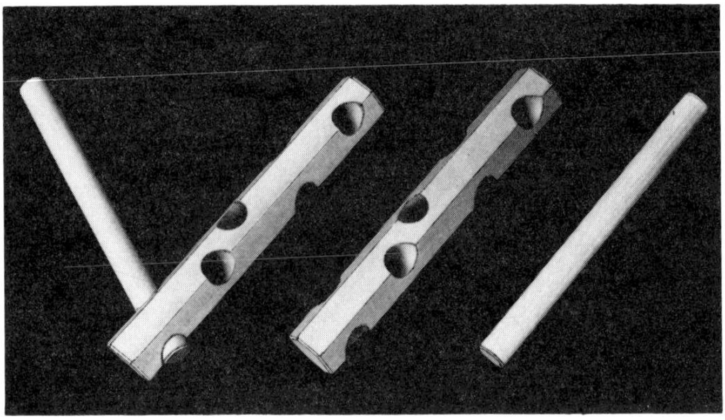

Fig. 129

Hexagonal sticks are easily made by first ripping planed boards into sticks of rhombic cross-section with the saw tilted 30 degrees and then making two more cuts. All of the holes are spaced equally apart, are at the same $70\frac{1}{2}$-degree angle to the axis of the stick, and are arranged in helical progression. Thus, a simple drilling set-up can be used that positions the stick using the previously drilled hole, with the stick being rotated 120 degrees in the same direction each time. The spacing of the holes can be determined by trial and error to achieve a snug fit. If they are too close together, the puzzle cannot be assembled. Spacing them farther apart simply makes a more open arrangement. With an open arrangement on a large scale, what a delightful and attractive climbing apparatus could be made for a children's playground.

This lattice structure repeats itself indefinitely in all directions, so one can make larger assemblies with more and longer sticks and dowels. From among the infinite variety of such constructions, one example is shown in Fig. 130a. It is basically two clusters joined together along their threefold axes.

Another fascinating feature of this construction is that sub-units are also possible using fewer and shorter sticks and dowels. From among the many possibilities, one example is shown in Fig. 130b. It uses four sticks and four dowels, and each stick has three holes. As an assembly puzzle it would be rather too easy if given the illustration of the solution. However, this is easily corrected by joining one stick-dowel pair to make an elbow piece and another pair to make a cross piece. This construction might also be used to make a novel collapsible stand for a tabletop.

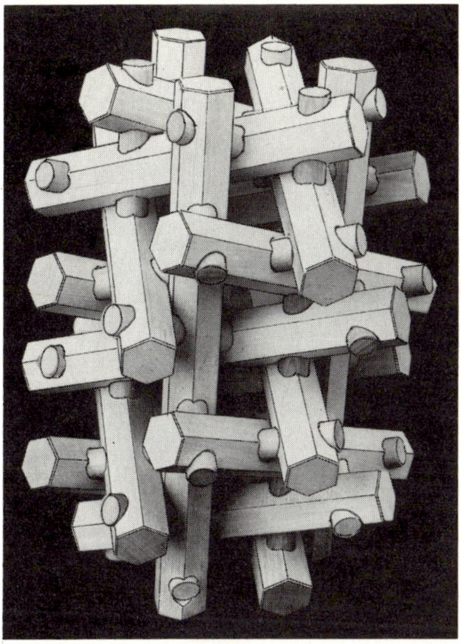

Fig. 130a

Fig. 130b

Yet another intriguing aspect of this system is its possiblities as a play construction set. Imagine having many sticks and dowels of each size from two-hole to five-hole and then discovering all the possible symmetrical constructions starting with the smallest and building upward. A few of these are shown in Fig. 131. What a marvellous plaything this might make for some curious youngster.

The Cuckoo Nest Puzzle

By making the arrangement of the holes alternate rather than helical, one obtains a different sort of lattice structure, which likewise can be extended indefinitely in all directions. Constructions made with it can have an axis of symmetry but not isometric symmetry. The version shown in Fig. 132 uses six sticks and six dowels, with each stick having three holes. It has a threefold axis of symmetry. If five stick-dowel pairs are joined together to make elbow pieces, it is a surprisingly difficult assembly puzzle with two solutions. Rather than show how the pieces are formed, we let the curious tinker enjoy the task of rediscovering them. Minor variations are possible, but there is no way to avoid having two pieces identical. A version of this puzzle was once produced under the name of *Cuckoo Nest.* By the way, note the functional similarities of this puzzle to the Pin-Hole Puzzle on page 70.

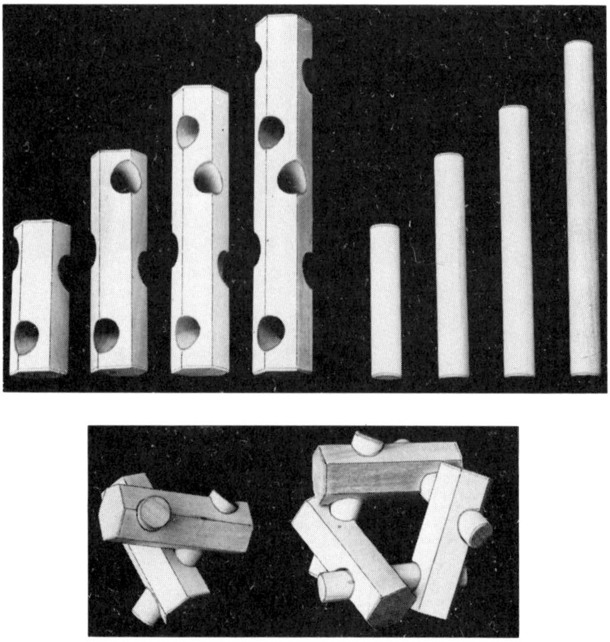

Fig. 131

Fig. 132

A triple decker puzzle

If the Cuckoo Nest Puzzle is regarded as having two layers, one way of expanding it is to add a third layer of three more sticks and three more dowels. One such version is shown in Figs. 133a and 133b. It likewise has a threefold axis of symmetry. It has only one solution and is more difficult than any of the above. Other even more complicated versions are possible with additional layers.

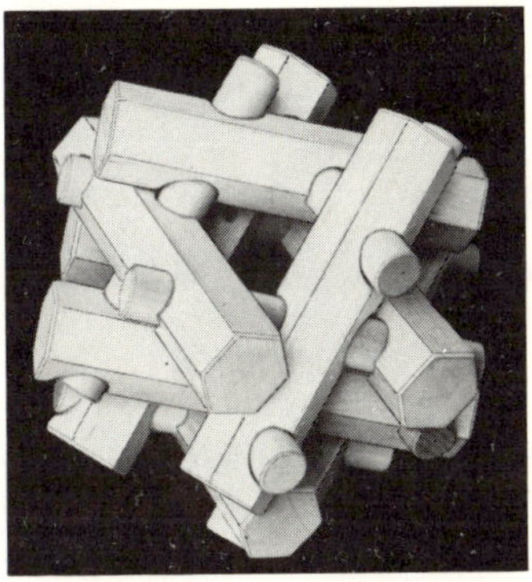

Fig. 133a

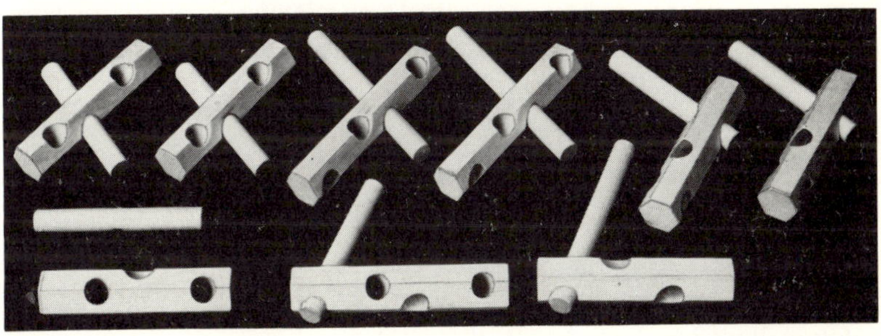

Fig. 133b

A holey hex hybrid

Never underestimate the amazing ways that geometrical constructions can be made to fit together in space. Just when we think we have exhausted the possible hole arrangements for symmetrical hexagonal stick assemblies, yet another one is discovered. Note the arrangement of the four holes in the hexagonal stick shown in Fig. 134. If the hole at the bottom is ignored, the arrangement is helical;

Fig. 134

but if the hole at the top is ignored, the arrangement is alternate. Eight such identical sticks, together with their eight corresponding dowels, can be assembled into a structure having one fourfold axis of symmetry and four twofold axes—i.e. the same as that of a square prism. Furthermore, if four stick-dowel pairs are now joined to make elbow pieces, it becomes a most perplexing assembly puzzle. The elbow pieces have two possible forms, thus providing further amusement and bafflement for the determined puzzle analyst.

Notched hexagonal sticks

The basic cluster of 12 triangular sticks, upon which most of the designs in this and the previous five chapters have been built, suggests the possibility of converting them into interlocking notched hexagonal sticks. With two trapezoidal notches in each stick, they form a neat interlocking structure of 12 identical pieces, as shown in Figs 135a and 135b, but which is impossible to assemble. A third notch in three of the pieces allows the puzzle to be assembled.

There are three distinctly different solutions to this puzzle, which can be defined by the arrangement of the three-notch pieces. The easiest and most obvious solution is with these three odd pieces going in last in a triangular arrangement to complete the assembly. In the second solution, the three odd pieces are mutually parallel, and there is a key piece that slides in last. The third solution is

Fig. 135a

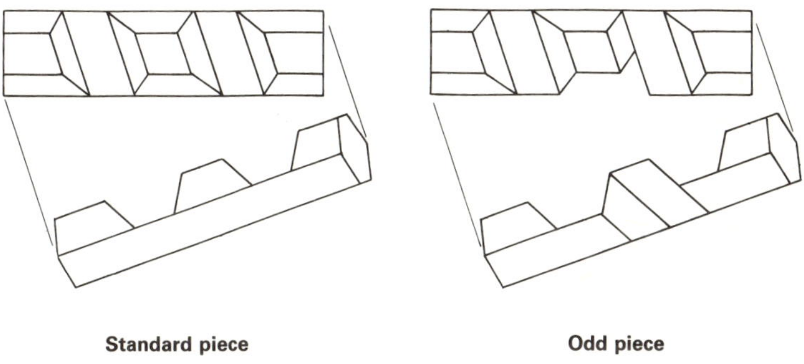

Fig. 135b

more difficult. Here is a case where multiple solutions make the puzzle more interesting.

This puzzle is enhanced by using four colours for the pieces, three of each colour. If the three odd pieces are the same colour, the first two solutions can have different sorts of colour symmetry. The second solution is especially interesting, with all like-coloured pieces being mutually parallel.

This puzzle was at one time manufactured in plastic as *Hectix*, but unfortunately never in four colours (saving the manufacturer a penny or two). A few have been produced in wood, which is quite easy with a supply of hexagonal stock and a trapezoidal cutter. Aside from its considerable potential as a very satisfactory assembly puzzle, it would make a handsome sculpture in brass or stainless steel, or perhaps even cast in concrete on a massive scale. (Reference: US Patent 3721448 to Coffin, 1973.)

Notched rhombic sticks

The puzzle described above leads by analogy to a cluster of 12 notched sticks having rhombic cross-section. They might also be regarded as augmented triangular sticks, especially since they are most easily fabricated by gluing triangular stick segments together. One possible version of such a puzzle is illustrated in Fig. 136, which uses three types of pieces, four of each. It has a rather unusual solution. Many other possible versions await investigation.

Fig. 136

Chapter 14

Split triangular sticks

Referring back once again to the now familiar cluster of 12 triangular sticks, divide each stick in two longitudinally. This produces a totally symmetrical arrangement of 24 sticks of 30–60–90-degree triangular cross-section. These sticks are then joined in fours to make a simple but intriguing geometrical puzzle. In the model shown in Fig. 137, the ends of the sticks have been cut off at an angle, giving the assembled puzzle the envelope of a rhombic dodecahedron with eight hexagonal dimples and six square dimples.

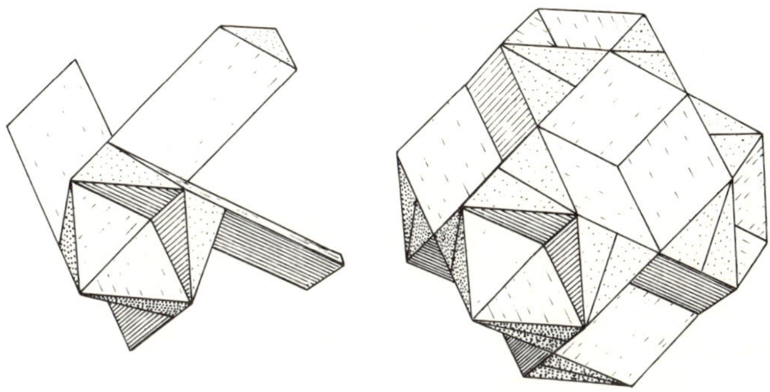

Fig. 137

When accurately made, this puzzle feels solid when handled, and it may take a while to discover that it can be slid apart along any one of its four sliding axes into two identical halves. The puzzle is assembled by the reverse of this with no great difficulty. However, the assembly sometimes has a surprise ending. For some reason, many persons feel the urge to toss this puzzle into the air after they have assembled it. Perhaps it is because it is more nearly spherical than most and feels so solid. The theory of interlock indicates this to be truly an interlocking configuration, but it fails to take into account that the pieces all rotate slightly and free themselves from one another. The usual result is that it flies apart in all directions!

This puzzle lends itself very well to multicolour problems. Four contrasting colours are used for the 24 sticks, six of each colour. The six puzzle pieces are made up of all possible permutations of the four colours. There are four solutions having colour symmetry. In the first and most obvious, like colours are mated in pairs and all like colours are mutually parallel. In the second, like colours are again mated in pairs but form four rings around the outside. In the third, all hexagonal dimples are solid colours. In the fourth, no like colours touch each other. The black and white representations of these are shown in Fig. 138. Discover a simple transformation from one to another.

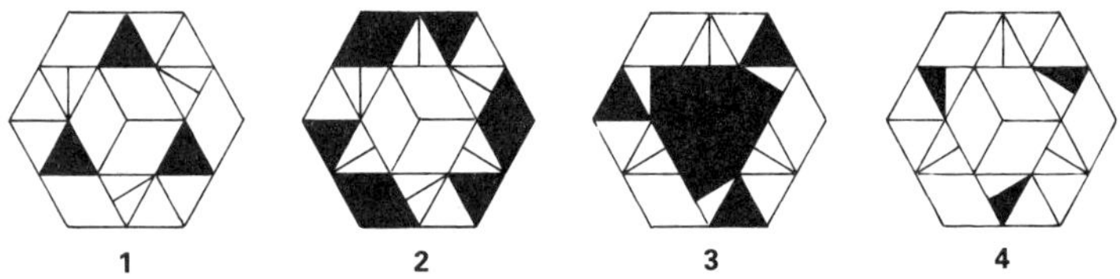

Fig. 138

This puzzle was produced at one time under the name of *Scorpius.* Since the configuration is very useful and leads to many other interesting designs, rather than refer back to it by some complicated yet ambiguous geometrical description, it will be more convenient simply to call it the Scorpius configuration.

The Dislocated Scorpius Puzzle

One interesting variation of the Scorpius Puzzle has six puzzle pieces that are identical but non-symmetrical. As shown in Fig. 139, this is accomplished by starting with the symmetrical Scorpius piece and exchanging one arm with its mating half, hence the name. This puzzle has two distinctly different mechanical solutions. If four different colours are used for the arms, there is one way of colouring them such that the two different mechanical solutions produce two different symmetrical colouring patterns, corresponding to the first two shown for the Scorpius Puzzle.

The Scrambled Scorpius Puzzle

The 24 sticks in the Scorpius configuration lend themselves naturally to being joined in fours in different ways to create a combinatorial puzzle. Not counting side-by-side pairs, there are 10

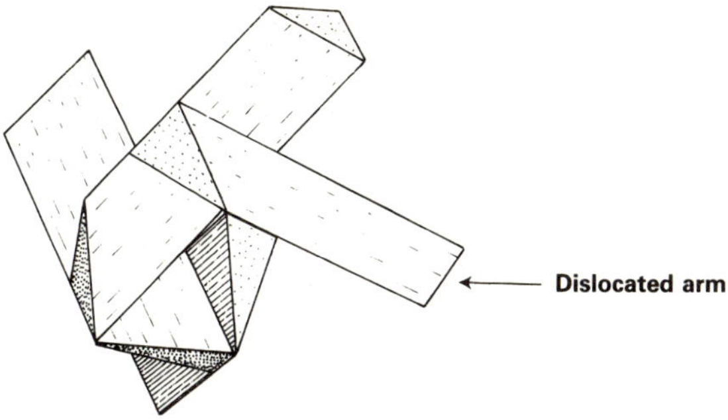

Fig. 139

different ways of joining four such sticks. Of these, one is symmetrical, two are impossible to assemble, and one does not permit any solutions. By a most extraordinary stroke of luck, the remaining six pieces, shown in Fig. 140, assemble one way only with only one sliding axis and in essentially only one possible order to create a combinatorial puzzle of intriguing geometry and considerable difficulty.

For whatever it may be worth, the difficulty index of this puzzle is calculated as follows: the six pieces can be regarded as positioned

Fig. 140

roughly as the faces of a cube—top, bottom, front, back, left, right. Arbitrarily choose any piece for the bottom. Next, the piece on the left can be any one of the remaining five, oriented any one of four ways, and so on. So, $20 \times 16 \times 12 \times 8 \times 4$ gives a difficulty index of 122 880 with just six simple pieces. Added to that is the amusement of figuring out the one correct order of assembly. All things considered, this is a most satisfactory puzzle and one that ought to be produced and enjoyed much more so than it has been. (Reference: US Patent Des. 230 288 to Coffin, 1974.)

Chapter 15

Dissected rhombic dodecahedra

On page 96, a rhombic dodecahedron was shown dissected into 12 rhombic pyramids. In the model shown in Fig. 141, each rhombic pyramid is further divided into two identical halves that could be regarded as skewed rhomboid pyramids or triangular stick segments. Not counting side-by-side pairs, there are 10 different ways of joining four such blocks together, analogous to those of the Scrambled Scorpius Puzzle. Nine of these are non-symmetrical and are illustrated in Fig. 141.

Problem: from this set of nine pieces, find subsets of six that assemble into the rhombic dodecahedron. Two such subsets are

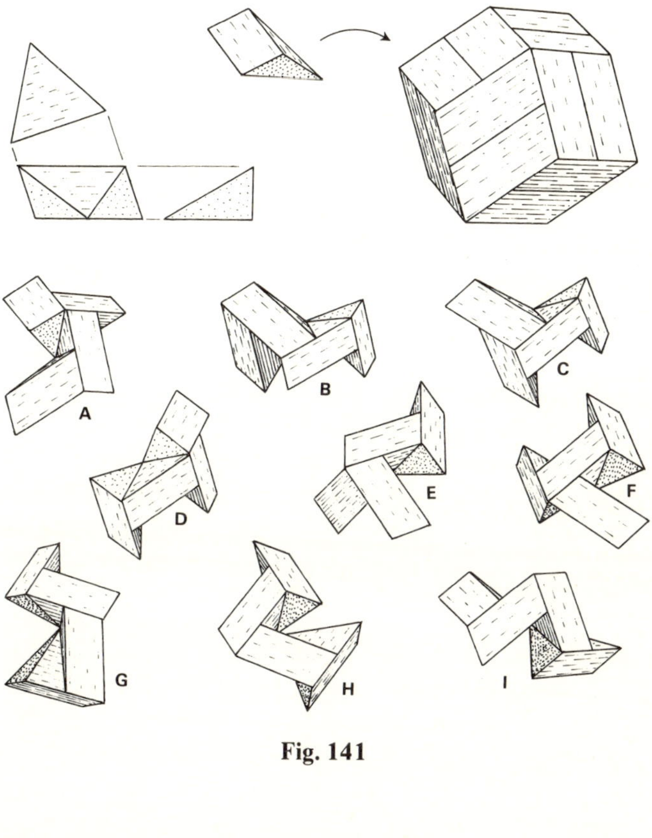

Fig. 141

known, and there are probably no others. Either subset makes a satisfactory interlocking puzzle with only one solution and one sliding axis. Five of the pieces are common to both subsets, so an especially interesting version of the puzzle is a set of seven pieces that will construct either solution with one piece set aside.

Since this is a fairly easy puzzle to make, the reader is encouraged to do so and discover these two solutions or perhaps experiment with other sizes of pieces and new combinations. The 24 blocks are sawn from 30–60–90-degree triangular cross-section sticks as shown in Fig. 142. If sawn accurately, they tend to align themselves properly when clustered together and held with rubber bands and tape. The desired joints are then glued selectively, one at a time. The finished puzzle, well waxed and with one piece removed, can then be used as a gluing jig for the next one.

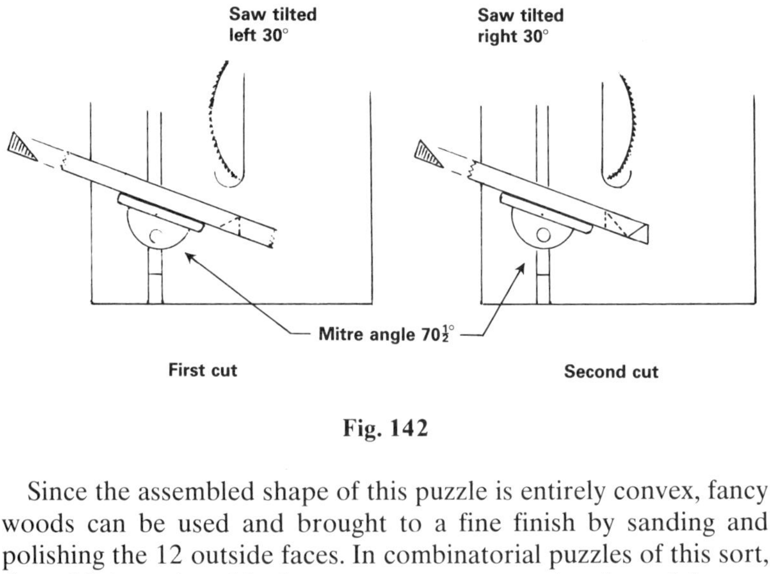

Fig. 142

Since the assembled shape of this puzzle is entirely convex, fancy woods can be used and brought to a fine finish by sanding and polishing the 12 outside faces. In combinatorial puzzles of this sort, the addition of multicolour symmetry to an already satisfactory puzzle tends to defeat its purpose. Instead, an attractive random mosaic effect is obtained by making each puzzle piece of a different wood in contrasting colours.

Note also that this is one of the few designs in which all the planes of dissection pass through the centre of the puzzle. Consequently, the assembled puzzle can be truncated or rounded down to various sculptural shapes (Fig. 143), making interesting and sometimes surprising patterns in the multicoloured versions. By the

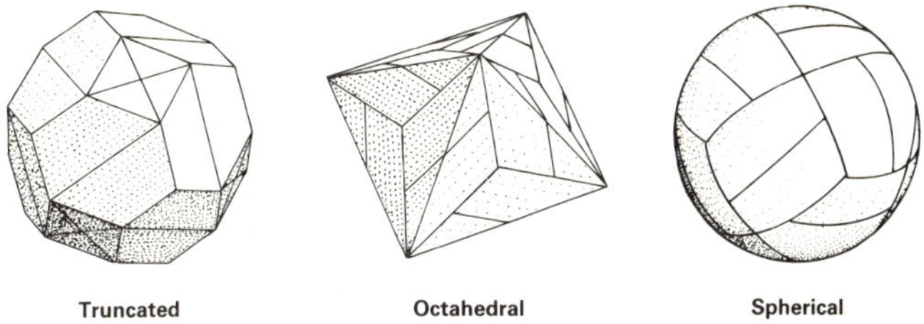

Fig. 143

same token, the assembled puzzle can be either solid or hollow inside.

Two-tiered puzzles

The 24-piece dissections of the rhombic dodecahedron and the Scorpius family of puzzles have many characteristics in common, including the fact that one fits exactly inside the hollow centre of the other. This suggests the intriguing possibility of a two-tiered puzzle construction. With 48 individual blocks or sticks to work with, the possible puzzle pieces and combinations are practically limitless. Just one example is shown in Figs 144a and 144b. It is made of 48 identical blocks and has the assembled shape of the first stellation of the rhombic dodecahedron, slightly truncated. It may violate the rule that the simplest designs are usually the best,

Fig. 144a Partially disassembled

Fig. 144b

but it does so in such glorious fashion, who cares? Perhaps someone will now be encouraged to try a three-tiered design!

The Pennyhedron Puzzle

Now to the other extreme! It has already been shown that polyhedral puzzles need not have many pieces to be interesting and even challenging. The confusing Three-Piece Block Puzzle speaks for itself. The fewest pieces an assembly puzzle can have is, by definition, two. Is it possible to create an interesting assembly puzzle of just two pieces?

When our three children were quite small, they used to spend hours in my workshop patiently gluing together little scraps of fancy woods to make 'puzzles' for their friends. One time we had a surplus of truncated rhombic pyramid blocks which they industriously glued together all different ways. What emerged from this was a simple two-piece dissection of the rhombic dodecahedron (Fig. 145). It has two mirror-image halves made of six blocks each that fit together with no difficulty whatsoever. It is when you try to take it apart that the fun begins. If made carefully so that the division of the two halves does not show, nearly everyone will grasp randomly with thumb and forefinger of each hand on opposite faces and pull. But when you do that, you will always be holding both pieces in each hand, and it will never come apart. Only when one uses an

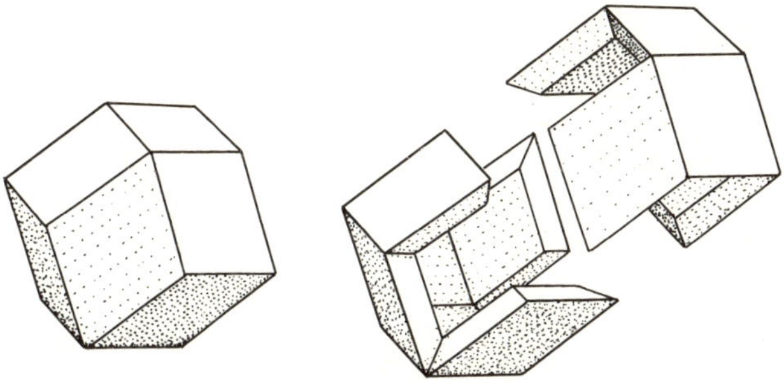

Fig. 145

unnatural three-finger grasp with each hand and then hunts randomly for the one sliding axis will it come apart with ease!

We made and sold these for a few years. The kids used to put a penny inside, hence the name. I think they used up all the scraps and got interested in other things at about the same time, and production ceased. We tinkered with many variations. The only limit here is your imagination. There are truncated and stellated versions, rounded, three-piece, and multicoloured ones. There are nesting sets in which each one is different. A few random samples are shown in Fig. 146.

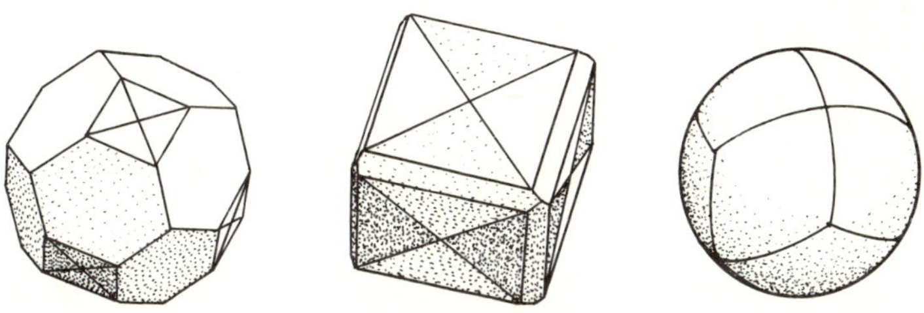

Fig. 146

One of the most amusing and confusing versions was a matched set of two Pennyhedrons—each one made of 24 tetrahedral blocks, as shown in Fig. 147. One of these is the standard Pennyhedron that comes apart with the tricky three-finger grasp. The other one, which looks exactly the same, comes apart easily along a fourfold axis of symmetry with the common thumb-and-forefinger grasp. Naturally the kids could do the tricky one, but the easy one had them completely baffled! (We will be seeing this dissection again in chapter 21.)

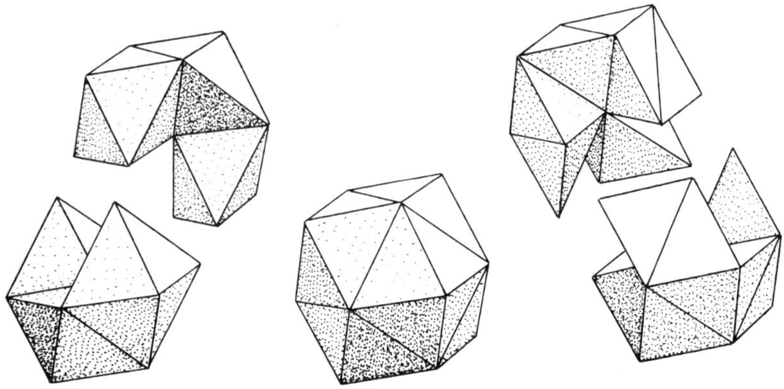

Fig. 147

Chapter 16

Miscellaneous confusing puzzles

Rightly so or not, the puzzle inventor is often perceived as a fiendish sort whose only purpose is to confuse and frustrate others. Witness the names frequently given to the instruments of the profession—The Devil's Dice, Instant Insanity, The Diabolical Cube, and so on. Anyone who has ever sold puzzles over the counter at craft shows has been asked many times for a puzzle that will drive someone else crazy (usually a close relative!). In this book, I have tried for the most part to present the other side of the coin—geometrical recreations that are fascinating and often challenging, but where confusion is not the ultimate object and deception is not the means to that end. The Pennyhedron Puzzle just described, especially the confusing pair, bears witness to good intentions gone astray. In this chapter, I have purged my files of a few other designs in the same deviant vein.

The Pseudo-Notched Sticks Puzzle

Readers familiar with the symmetrical version of the diagonal burr puzzle know that the easiest way to disassemble it, especially when it is tight, is to grasp any opposite pair of sticks, wiggle and pull, and the whole thing flies apart. The Pseudo-Notched Sticks Puzzle (Fig. 148) also has six identical pieces and looks exactly like the diagonal burr when assembled. But when you grasp what appear to be two opposite sticks to disassemble it, all you are doing is pressing it ever more tightly together, and it has the feel of being glued absolutely solid! Only when one grasps in a manner that seems to make no sense at all does it come apart with ease!

Anyone who collects puzzles or writes about them is faced with the question of classification. This design illustrates the sort of problems one confronts. What other field of human endeavour outside of the legal profession is so purposely confusing? If based on superficial appearance, this puzzle would be in chapter 7, but psychologically it belongs here.

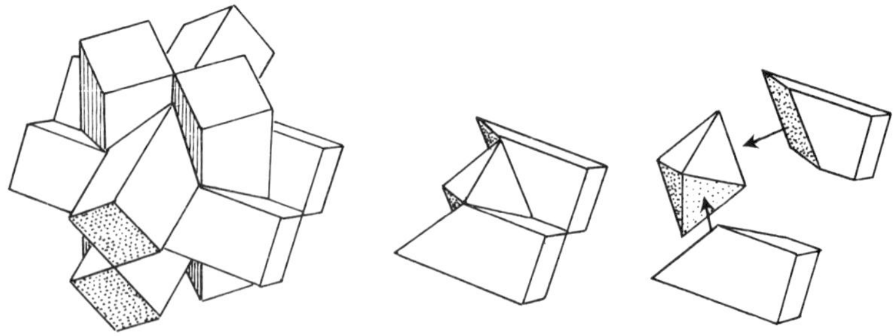

Fig. 148

The Square Face Puzzle

The puzzle shown in Fig. 149 is made by starting with the Pseudo-Notched Sticks Puzzle and adding 12 more notched blocks. Some minor variations are possible, depending upon just where the blocks are attached. The version shown here has six dissimilar, non-symmetrical pieces and two solutions. Note that the addition of the 12 blocks completely changes the appearance of the puzzle so that it now more nearly resembles the Three Pairs Puzzle, even though functionally those two are entirely different.

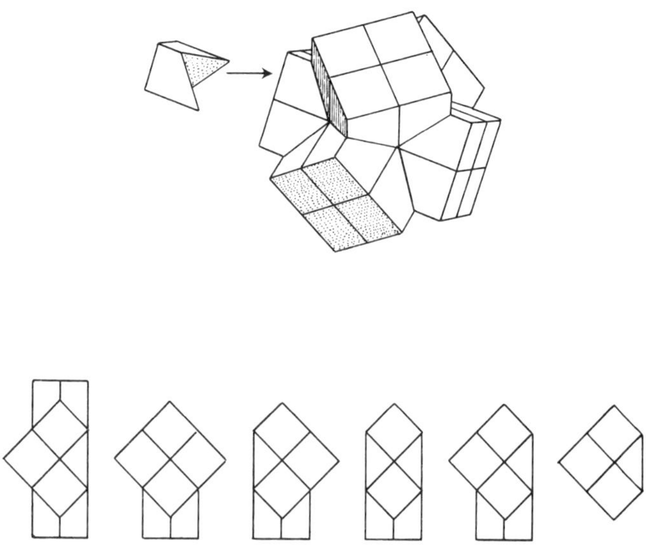

Fig. 149

The Queer Gear

Readers are probably familiar with the novelty consisting of two blocks joined together by what looks like a pair of dovetail joints impossible to assemble or disassemble, see Fig. 150. Deceptions of this sort sometimes appear in polyhedral puzzles. One does not expect the Star of David Puzzle to come apart along a diagonal sliding axis. This principle is exploited in the design shown, which has a perfectly prismatic assembled shape but in which the first step of disassembly is a separation into two halves along an unexpected diagonal axis. The construction of these puzzle pieces is similar to that used for the intersecting prism puzzles in chapter 10, with triangular stick segments attached to six-sided centre blocks. The six dissimilar pieces of the version shown are in reflexive pairs. The two end faces are finished true after assembly.

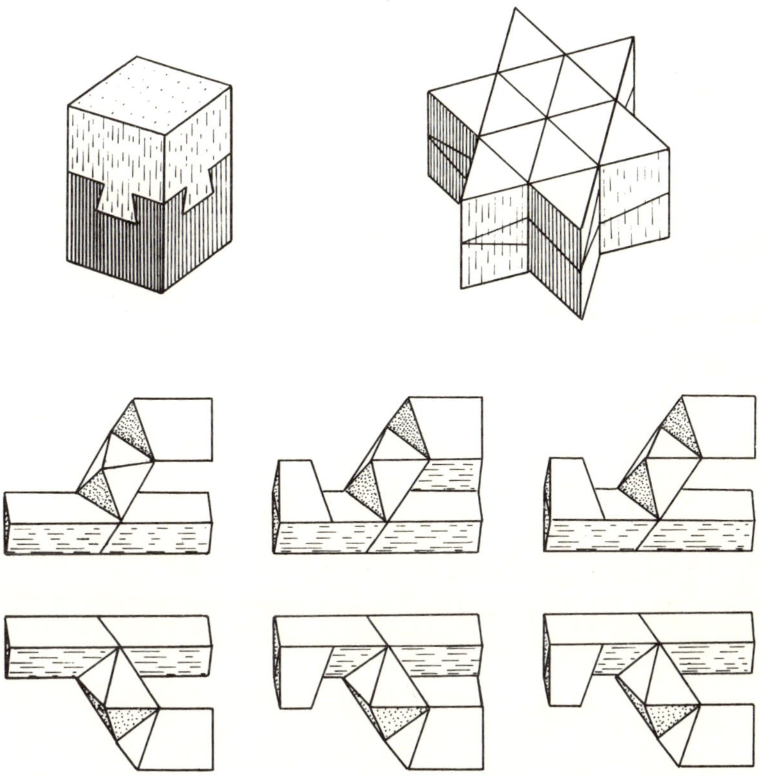

Fig. 150

Note that none of the designs described in this book, even in this diabolical chapter, employs concealed locking devices such as hooks, catches, tumblers, or the like. Patent files reveal that one of the preoccupations of puzzle inventors over the years has been to devise such mechanisms, the object being to defeat them and open some secret box or toy bank. Does this fascination with locks and concealment tell us something about ourselves? The number of security devices that one must necessarily deal with in daily life is one of the most discouraging aspects of our civilization. Yet how casually we take them for granted, even turning them into recreations and children's toys!

Chapter 17

Triacontahedral designs

Besides the cube and the rhombic dodecahedron, the only other polyhedron that can be totally enclosed by a symmetrical arrangement of sticks is the 30-faced triacontahedron (Fig. 151).

An obvious approach to exploring the geometry of the triacontahedron for practical applications is to refer back to the previous chapters in which the mating surfaces of the puzzle pieces corresponded to faces of the rhombic dodecahedron and see which designs can be carried forward by analogy into this new geometry. Straight away, one finds that there is nothing equivalent to the diagonal burr in triacontahedral geometry, so none of the designs described in chapters 7 through 12 have triacontahedral offspring.

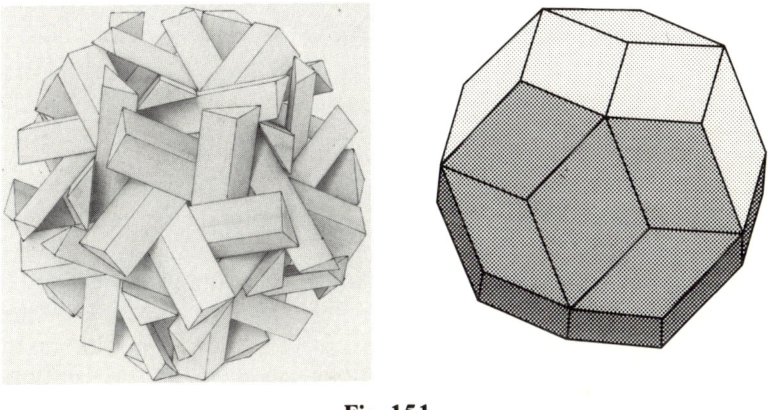

Fig. 151

Thirty pentagonal sticks and dowels

The cluster of 12 hexagonal sticks and dowels shown in Fig. 128 has an analogy in 30 pentagonal sticks and dowels. The version shown in Fig. 152 has seven holes in each stick. A smaller version is possible using shorter sticks and dowels, with five holes in each stick; while yet another smaller and more spherical version has only three holes in each stick. There is also one larger and more stellated version with nine holes in each stick.

Fig. 152

The construction can be extended along any one axis and the structure will repeat itself in an indefinite chain. It will not form a three-dimensional space-filling crystallographic lattice, however—none are possible having fivefold symmetries.

Pentagonal sticks are easy to make by first ripping $\frac{3}{4}$-inch boards into trapezoidal sticks and then making two more cuts, all with the saw tilted 18 degrees (Fig. 153). All of the holes are drilled at an angle of $63\frac{1}{2}$ degrees to the axis of the stick, passing through the centre of the stick, and parallel to one face. Determining their irregular spacing is the tricky part. It could be calculated, but the author must confess he found it simpler to locate them by trial and error. Slight inaccuracies in the 210 holes can be corrected by reaming them through in the assembled or partially assembled state using a round file in an electric drill.

Assembly is entertaining and not too difficult if aided by an illustration. Of course, the use of elbow pieces as in the hexagonal counterpart would turn it into an exceedingly difficult puzzle. One can spend hours studying the assembled structure, pondering its many mysterious properties and admiring its beautiful symmetries.

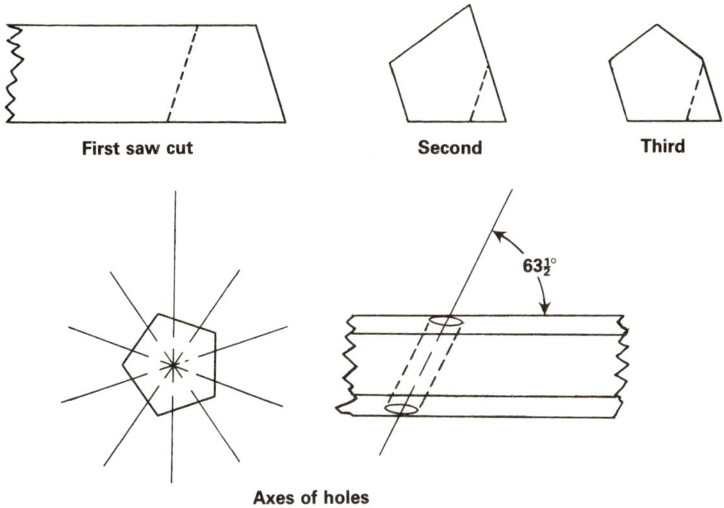

Fig. 153

Pentagonal sub-units

In chapter 13, a symmetrical assembly of 12 hexagonal sticks and dowels was broken into various sub-units with fewer and shorter sticks and dowels. By the same token, the assembly of 30 pentagonal sticks and dowels can be broken into interesting sub-units. One such is shown in Fig. 154 using five identical sticks and

Fig. 154

five dowels. Each stick has four holes. The assembly has fivefold symmetry. One puzzling version of it uses two elbow pieces.

Note the interesting genealogy of the above offspring. It represents the conjugation of two distinctly different ideas—the pentagonal geometry of this chapter and the sub-unit scheme of chapter 13, each with its own separate line of development, surprisingly and fortuitously joining neatly together. This happens all the time in the field of geometrical dissections and is just one more reason why this recreation is so fascinating.

Notched pentagonal sticks

The intriguing geometry of the Hectix Puzzle (Fig. 135) with its 12 notched hexagonal sticks suggests by analogy a cluster of 30 notched pentagonal sticks. Two versions of such a design are shown in Fig. 155, the difference between the two being a 36-degree rotation of the sticks. The model on the left is a mock-up only, as the sticks are so completely interlocked that any notching scheme to permit assembly would appear to cut some sticks completely in two. The complicated notching scheme that permitted assembly of the model on the right many years ago has long since been forgotten, as it has never been disassembled and serves only as a fascinating sculpture.

Fig. 155

Notched rhombic sticks

A more practical assembly puzzle is one made with sticks of rhombic rather than pentagonal cross-section. The rhombic sticks are easily made on a table saw with the saw tilted 18 degrees. The basic piece with two notches is shown in Fig. 156. Thirty such identical sticks are altogether impossible to assemble. The various schemes for modifying some of the pieces to permit assembly are left for the reader to invent. Careful inspection of the model shown should reveal the scheme used by the author.

In a similar vein, Bill Cutler has designed and constructed a puzzle made up of 30 notched *square* sticks. It is illustrated on p. 62 of *Puzzles Old and New.* Other variations are no doubt possible.

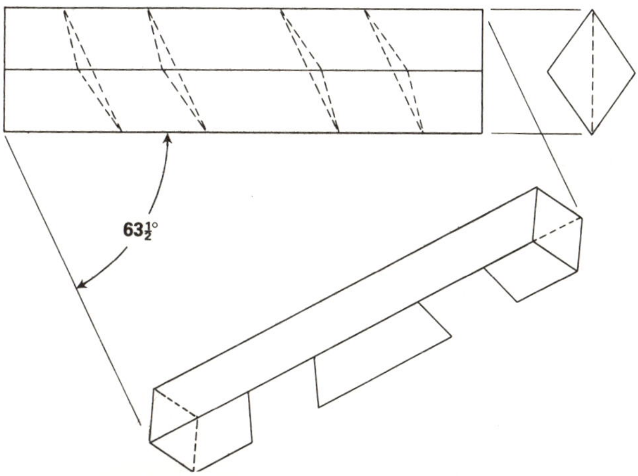

Fig. 156

The Jupiter Puzzle

The triacontahedron can be completely enclosed by an arrangement of 30 sticks of 36–108–36-degree triangular cross-section, as was shown in Fig. 151. If these triangular sticks are split longitudinally into two identical halves and then joined in fives to make 12 identical, symmetrical puzzle pieces, an interlocking configuration is obtained that is directly analogous to the Scorpius Puzzle. It has six sliding axes, and the final step of assembly is the mating of two identical halves. This puzzle likewise has the tendency to fly apart when tossed into the air, even more so than the Scorpius. In the model shown in Fig. 157, the ends of the sticks are trimmed at an angle, giving it the appearance of a stellated triacontahedron with 30 faces. One puzzle piece is also shown alone.

Fig. 157

Note that for all of the polyhedral design shapes included in this book, only one view of the assembly need be shown. One just naturally and automatically assumes that the structure is symmetrical, so any additional views would only be redundant. This assumption of symmetry, consistency, congruence, repetition, predictability, or call it what you will, is so commonplace, not only in geometrical recreations but in all the arts and sciences, that we scarcely give it a thought. Yet where would we be without it!

This design, as illustrated in Fig. 157, lends itself well to colour symmetry problems. Six colours are used, ten sticks of each colour. Each puzzle piece has arms of five different colours, arranged in such a way that when correctly assembled, all like-coloured sticks are in mutually parallel matched pairs, as indicated by the black and white representation in Fig. 158. Four other solutions having colour symmetry are then also possible, with a simple transformation from one to another. When well crafted of six dissimilar exotic woods, it is a fine specimen of woodcraft as well as a handsome geometrical sculpture. A puzzle of this sort was produced at one time with the name of *Jupiter*, and so for convenience it will be referred to by that name in what follows. (Reference: US Patent Des. 232 571 to Coffin, 1974.)

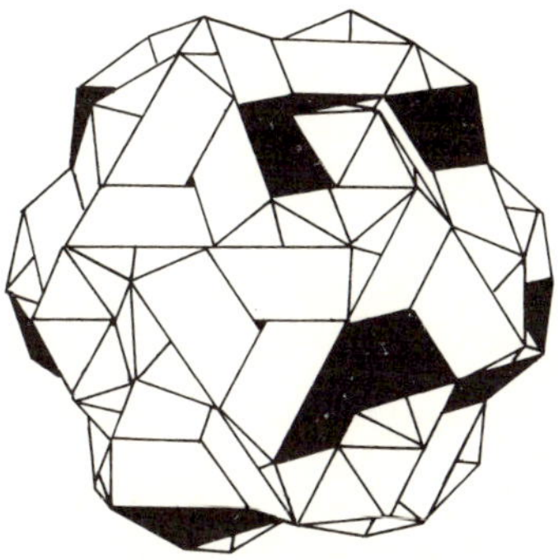

Fig. 158

A favourite theme of puzzle inventors is a device that looks deceptively simple to assemble but is actually quite difficult. The Jupiter Puzzle is an amusing example of just the contrary. Most persons will not even attempt to disassemble and reassemble this intriguing polyhedral dissection, so forbidding it looks; yet it is really quite easy. Years ago, when we worked the rounds of the craft fairs, I used to have one of these puzzles as the centre-piece of our display. When a crowd had gathered, I would toss it gently so that the pieces all fell in a heap. Then I would announce: 'Anyone who can put it back together can have it!' Usually no one would try, especially not adults. Our youngest, then about age seven, would be planted in the crowd, and you can guess the rest. As she finished it with a bored look, tucked it under her arm and walked off, the crowd would finally realize that it had all been just a joke on them!

The Dislocated Jupiter Puzzle

The Dislocated Scorpius Puzzle has a direct analogy in the Dislocated Jupiter Puzzle, with one arm of each puzzle piece being exchanged with its mating half. This puzzle is more complicated than the Jupiter. It has at least two distinctly different mechanical solutions. Colour symmetry problems are also possible. Some of the details of these remain to be worked out. One of the 12 identical puzzle pieces is shown in Fig. 159.

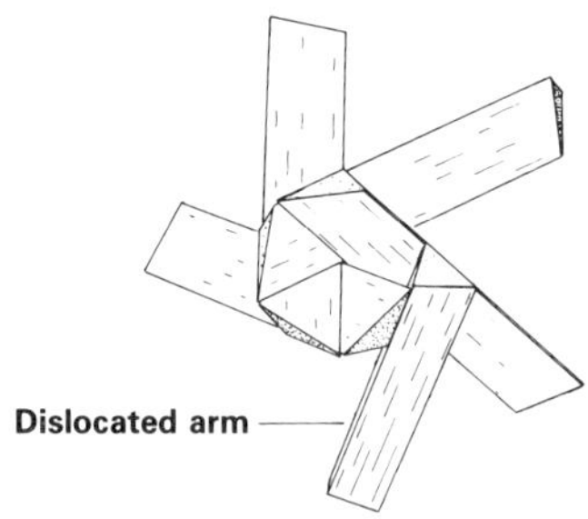

Fig. 159

A scrambled Jupiter?

The Scrambled Scorpius Puzzle, with its six dissimilar non-symmetrical pieces and one solution, suggests by analogy a triacontahedral version with 12 pieces. No such design has yet been found and may not exist. If it did, it would probably be too difficult to assemble without instructions. The compromise version shown in Fig. 160 has two identical halves of six pieces each. The six pairs of puzzle pieces are all dissimilar and non-symmetrical. The puzzle has multiple solutions. It is possible to use multicoloured pieces such that there is only one solution with colour symmetry.

The triangular sticks used in the Jupiter family of puzzles are of 36–54–90-degree cross-section. For the puzzle to be satisfactory, they must be glued together very accurately, using a gluing jig with the same angles as the vertex of a triacontahedron. Probably the only practical way to make the base for such a jig is on a milling machine. The ends of the sticks should not be any longer than shown (Fig. 161) or there will be interference and the puzzle cannot easily be assembled.

The dissected triacontahedron

The 24-piece dissection of the rhombic dodecahedron in chapter 15 leads by analogy to a 60-piece dissection of the triacontahedron,

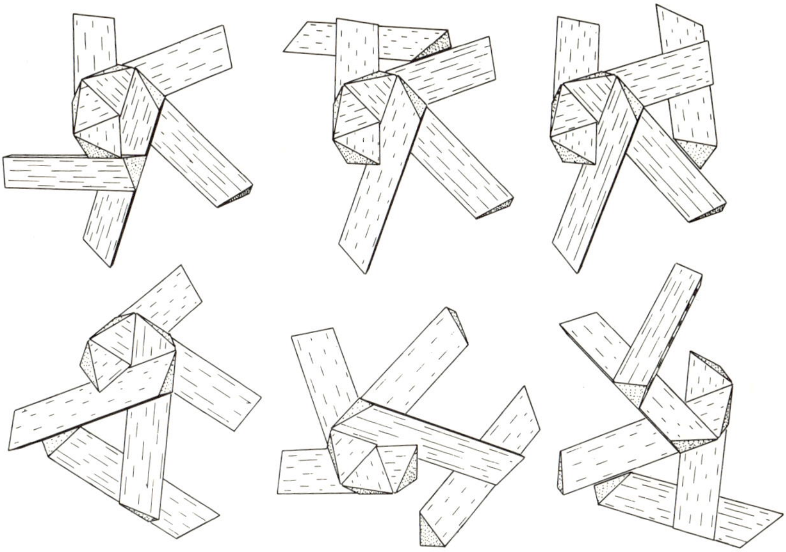

Fig. 160 The six Saturn pieces; two of each required

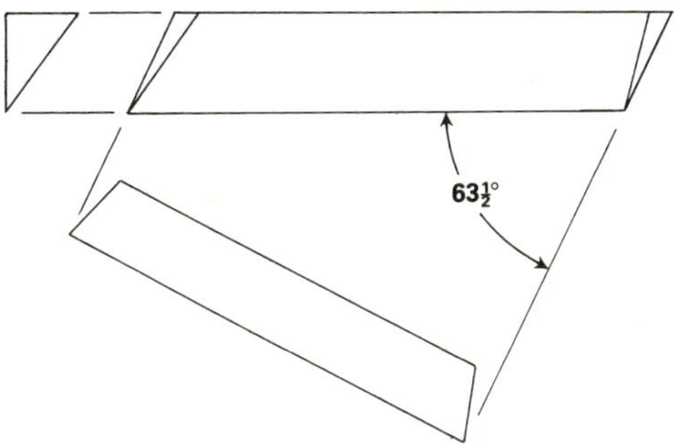

Fig. 161

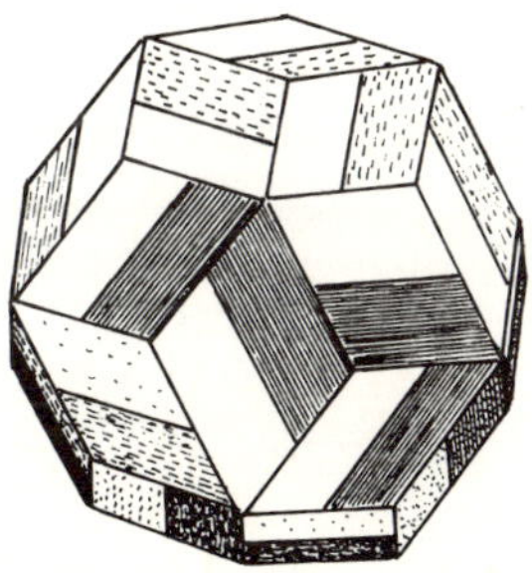

Fig. 162

as shown in Fig. 162. John Loeser has arrived at the data given in Table 5 for the possible ways of joining such blocks into different puzzle pieces, up to size-six.

Table 5

Size	Number of pieces	Total number of blocks
2	2	4
3	5	15
4	10	40
5	24	120
6	54	324

Of the 54 pieces of size-six, 45 of them are considered by us to be practical usable pieces. From among those 45, we have searched long and hard for a practical assemblable subset of 10 such pieces. So far, none has been found and it appears that perhaps none exists. But with some 20 billion different ways of choosing subsets of 10 from a set of 45, the possibility exists that we may have overlooked one! The example shown in Fig. 162 is randomly dissected simply to illustrate the principle rather than any practical solution.

Like those of the rhombic dodecahedron in chapter 15, these individual blocks are fairly easy to make. They are sawn from 18–72–90-degree triangular cross-section stock using the same techniques. Much wood and some labour can be saved by using trapezoidal rather than triangular stock, as shown in Fig. 163, thus making the centre of the puzzle hollow. The blocks are fairly easy to assemble and glue using tape and rubber bands to hold them in place. Considering the complexities and the failure to find a practical design, though, the potential for this dissection probably is not as an assembly puzzle but more as a geometrical sculpture, especially using multicoloured exotic woods.

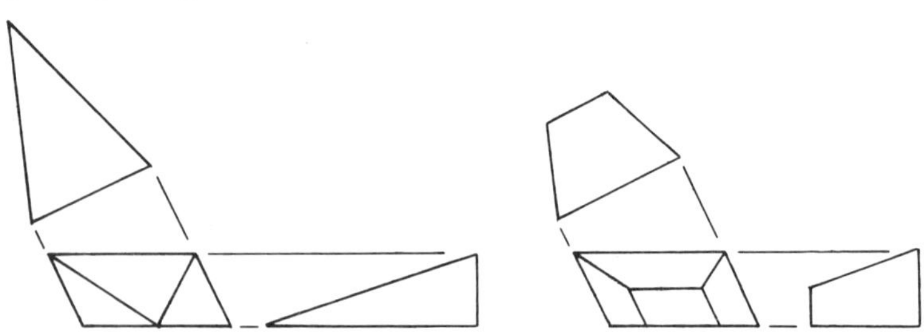

Fig. 163

Chapter 18

Puzzles made of polyhedral blocks

The ways in which various geometrical solids can be packed to fill space or be assembled symmetrically in space is in itself a fascinating branch of mathematics. Being also assemblable as puzzles just makes them all the more interesting.

Truncated octahedra

Besides the cube, there are only two other space-filling polyhedra that would appear to have much practical use as puzzle building blocks. One of these is the truncated octahedron. The blocks are fairly easy to make by starting with large wooden cubes and sawing off the eight corners using a suitable jig to hold them accurately (and safely!) in the saw (see Fig. 164). To check for accuracy, the eight new faces should be regular hexagons.

There are two ways that two such blocks can be joined and six ways that three can be joined, as shown in Fig. 165. These puzzle pieces tend to be anything but interlocking, so most practical puzzle constructions using them are either pyramidal piles or are packed inside a box.

The set of five pieces shown in Fig. 166 consists of two blocks joined by their square faces plus the four non-straight three-block

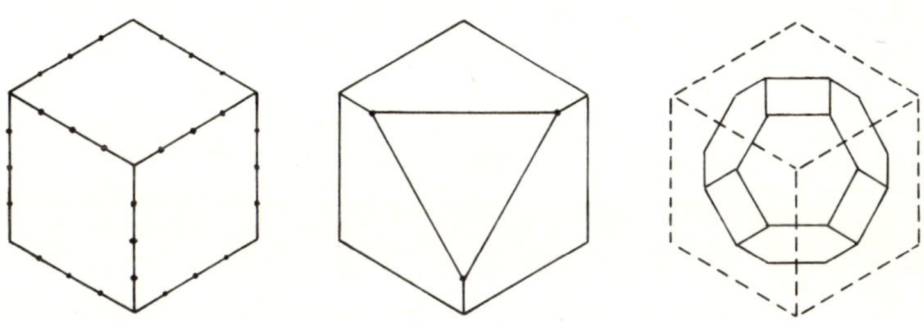

Dots indicate edges of cube divided into fourths | **First corner removed** | **Note hexagonal face**

Fig. 164

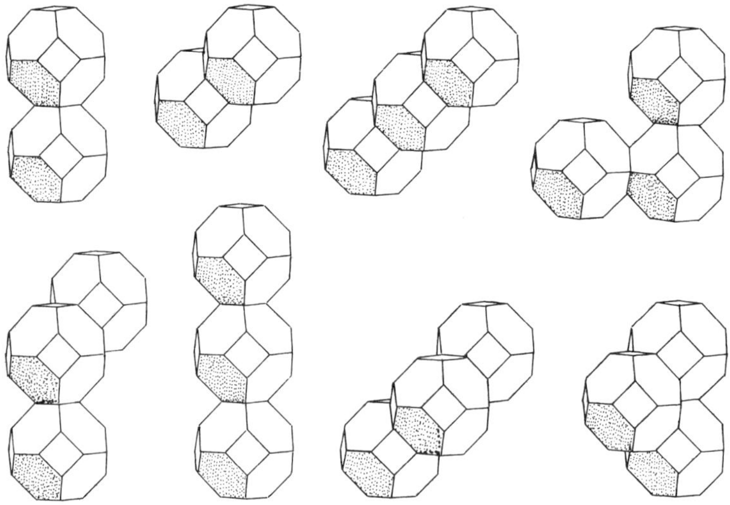

Fig. 165

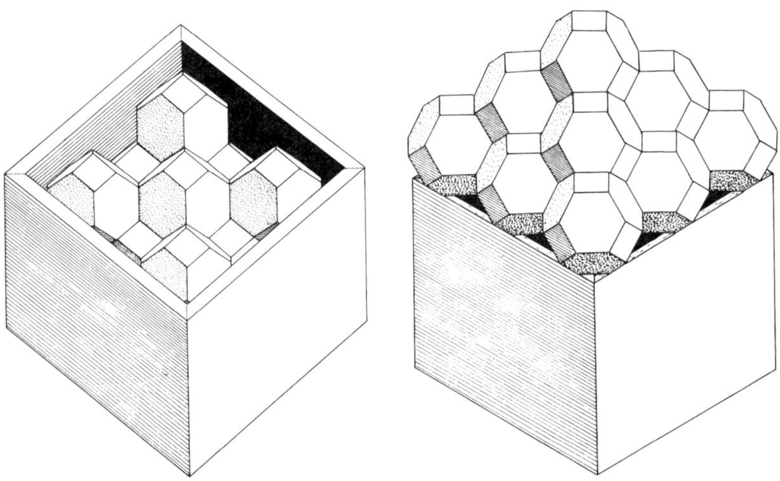

Fig. 166

pieces. They pack into the square box 11 different ways and make a square pyramid three different ways. Note that in the model shown the recessed bottom of the inverted box serves as a convenient base for the square pyramid. The instruction booklet that came with a commercial version of this puzzle showed 15 other symmetrical constructions using some or all of the pieces. How many can the reader discover?

Rhombic dodecahedra

Puzzle pieces made up of rhombic dodecahedra joined together different ways are fascinating to play with and offer practically unlimited possibilities for geometrical puzzle constructions and mathematical analysis. The blocks are fairly easy to saw from square stock using a special jig, described in chapter 23.

There is one way that two rhombic-dodecahedral (R-D) blocks can be joined, five ways that three can be joined, and 28 ways that four can be joined, as shown in Fig. 167.

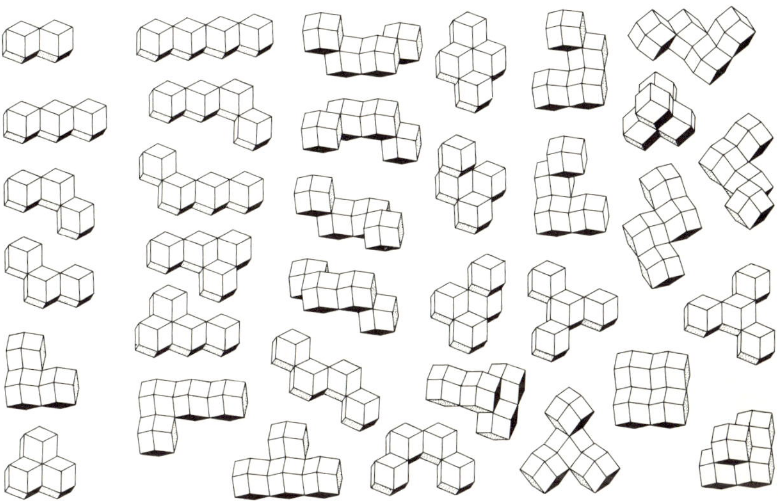

Fig. 167

An interesting exercise is to catalogue the practical geometrical constructions that might be possible using R-D blocks. All of those shown in Fig. 168 have isometric symmetry. They are arranged by increasing size, starting at the top. Those in the left-hand column have four blocks coming together at the centre. Those in the middle column have six blocks coming together at the centre. Those in the right-hand column have a single block in the centre, and hollow versions of these are possible using one less block. Only those constructions using 20 or fewer blocks are shown. It is convenient to identify these constructions by names such as *tetrahedron*, even though the shape is obviously not an exact tetrahedron but only suggests it using a little imagination.

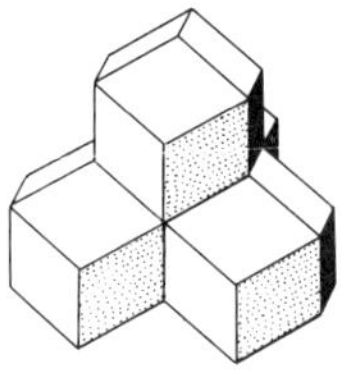

4-tetrahedron

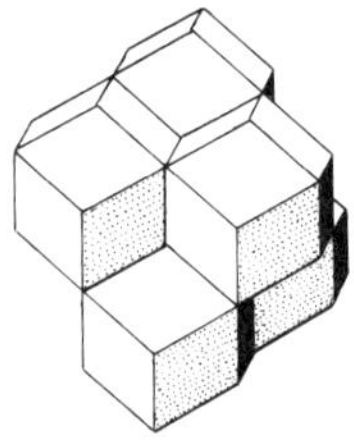

6-octahedron

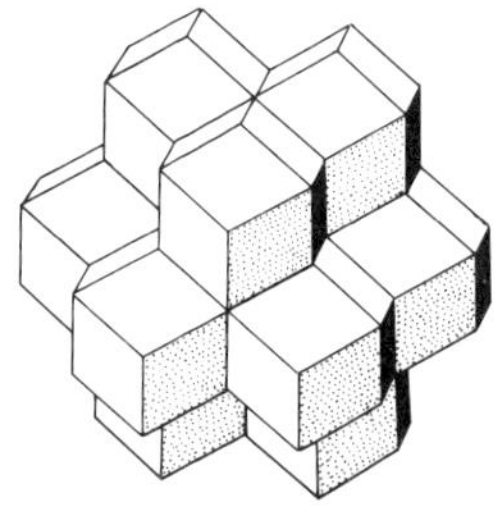

13-truncated octahedron

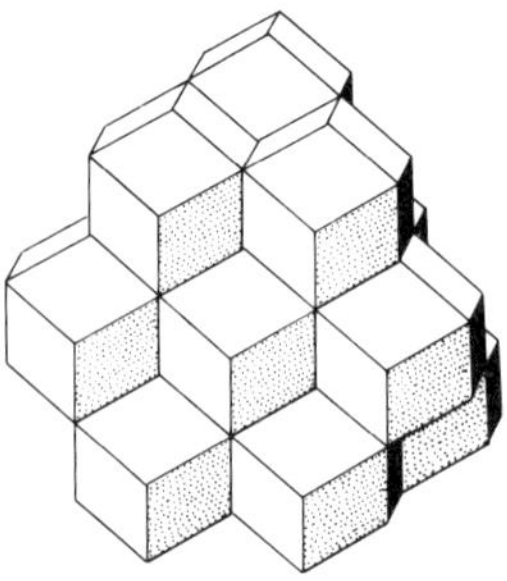

16-truncated tetrahedron

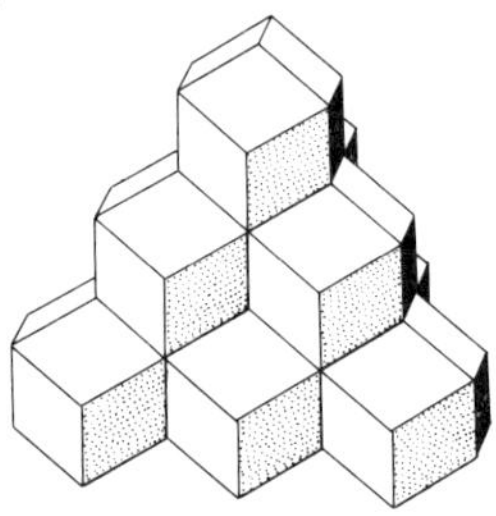

10-tetrahedron

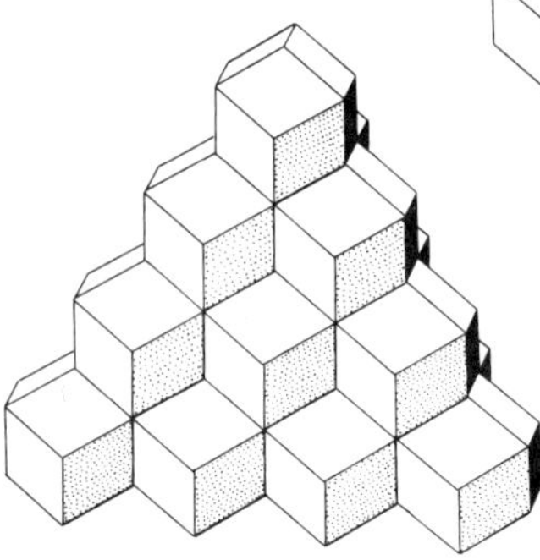

20-tetrahedron

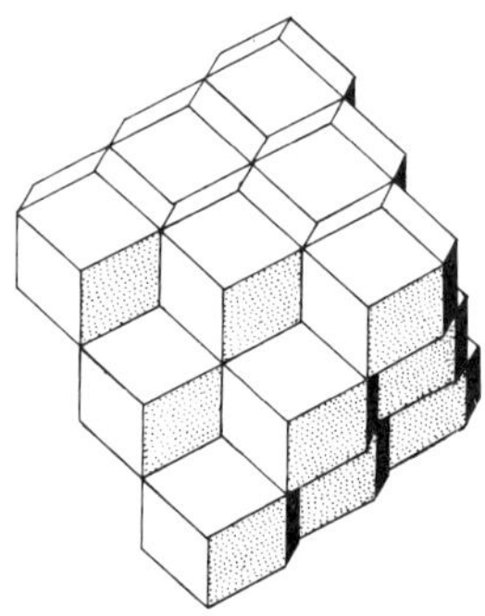

19-octahedron

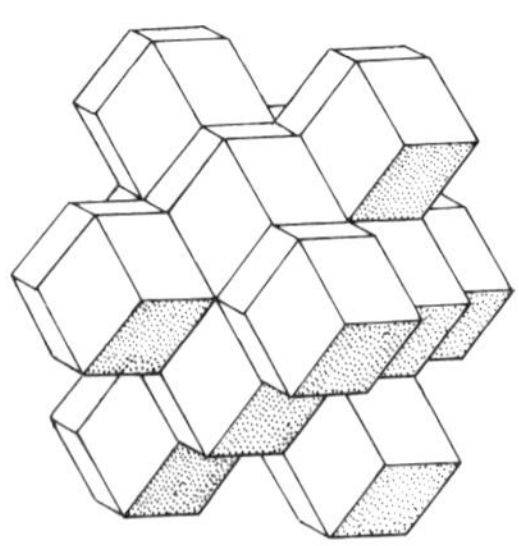

14-cube

Fig. 168

At this point, one has a choice of many possible avenues of exploration. What are the fewest pieces that will construct all of these shapes, or most of them, or other shapes of your choice? To put it another way, for a given number of pieces—say four or five or six—find the most versatile possible set that will construct many figures. Which sets have all dissimilar non-symmetrical pieces? Which figures have unique solutions? The recreational possibilities here are practically unlimited and largely unexplored.

The Leftover Block Puzzle

Polyhedral puzzles that are non-interlocking are usually more satisfactory if contained inside a box of some sort. Unlike cubes or even truncated octahedra, rhombic dodecahedra do not rest comfortably in square boxes. This can be corrected somewhat by truncating the R-D blocks. Shown in Fig. 169 is a simple but entertaining puzzle made up of 14 truncated R-D blocks joined together to make five puzzle pieces. They will construct a square pyramidal pile and also

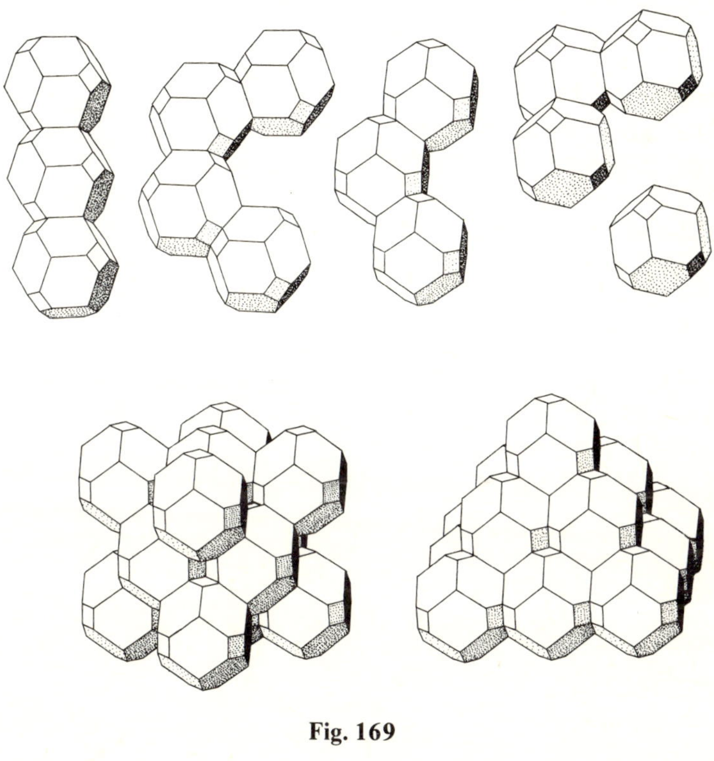

Fig. 169

a rectangular pyramidal pile. With one piece omitted, they will construct a tetrahedral shape that, surprisingly, fits neatly inside a box. All five pieces pack neatly inside a cubic box, as shown. But the most entertaining trick they do is pack snugly into the box apparently with no vacancy, but with the single block left out!

The easiest way to make truncated rhombic dodecahedra is to start with large cubic blocks and bevel all the edges at 45 degrees to any desired but uniform depth. With a shallow bevel, the mating surfaces will be small and the glue joints less strong, so it may be desirable to strengthen them by inserting dowels.

Substitution of spheres

Among the various ways that uniform spheres can be packed in space, they show a natural inclination to arrange themselves most densely the same way that rhombic dodecahedra pack. Thus, in the puzzle described above or almost any other, spheres might be substituted for rhombic dodecahedra. One advantage of spheres is that they are readily available in toy stores and educational supply shops, and usually quite accurate. The disadvantage is that they are more difficult to join together strongly. They can be bonded with epoxy, but an even better way with wooden balls is to drill holes and use dowelled joints.

The substitution of spheres for rhombic dodecahedra is not exactly equivalent mathematically. Spheres have an additional symmetry that the rhombic dodecahedra lack. This is demonstrated by the mirror-image pair of R-D pieces shown in Fig. 170, both of which have the same spherical counterpart. Thus, pieces made with spheres will generally produce more solutions and construct more figures, which could be an advantage or disadvantage depending upon the circumstances. Spherical versions also tend to fall apart more easily, so the pyramidal constructions may require a retaining base.

Of course, spheres might also be substituted anywhere that cubes

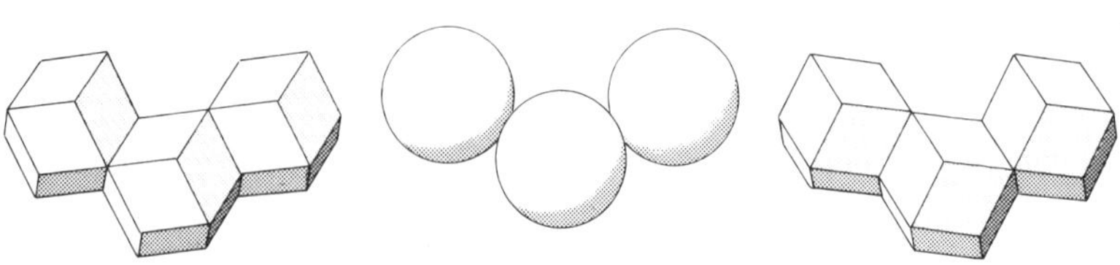

Fig. 170

are used. But like playing billiards with elliptical balls, one should ask not if it is possible but rather if it is *practical*!

The Four-Piece Pyramid Puzzle

Most polyhedral block puzzles are non-interlocking, including the two described above. All things considered, interlocking puzzles usually have more appeal. The problem here is one of complexity. For example, most practical, interlocking cubic-block puzzles require at least 64 blocks. The numbers of R-D blocks in triangular pyramidal piles are given by the following series: 4, 10, 20, 35, 56, 84, . . . which is the smallest of these that can be dissected into a practical *interlocking* puzzle? Surprisingly, the 20-block tetrahedral pile of rhombic dodecahedra can be dissected into four puzzle pieces of five blocks each that are not only interlocking but also dissimilar and non-symmetrical and assemblable in one order only. Knowing this, it is not very difficult to discover the design, so that recreation is left for the reader to enjoy. The one known design is believed to be unique but has not been proven so.

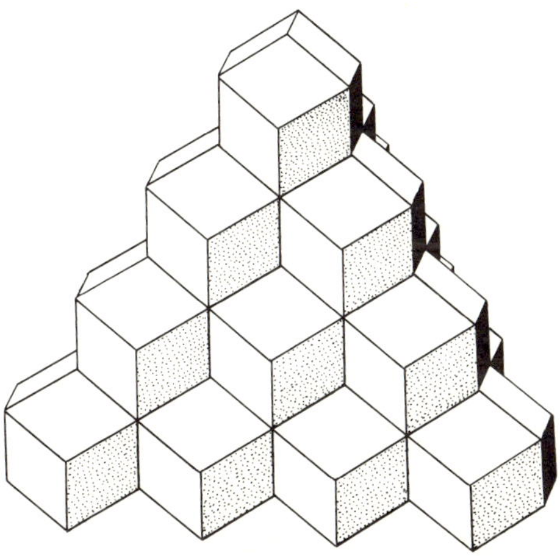

Fig. 171

The Octahedral Cluster Puzzle

The numbers of R-D blocks arranged in octahedral clusters are given by the following series: 6, 19, 44, 85, . . . It is especially desirable that a dissection of the octahedral cluster be interlocking because it would fit so poorly into a box. There is a four-piece dissection of the 19-block octahedral cluster that is interlocking and assembles in one order only. All of the pieces are dissimilar and non-symmetrical. Three of them have five blocks and one piece has four blocks. Again, its discovery is left to the reader. The one known dissection may or may not be unique.

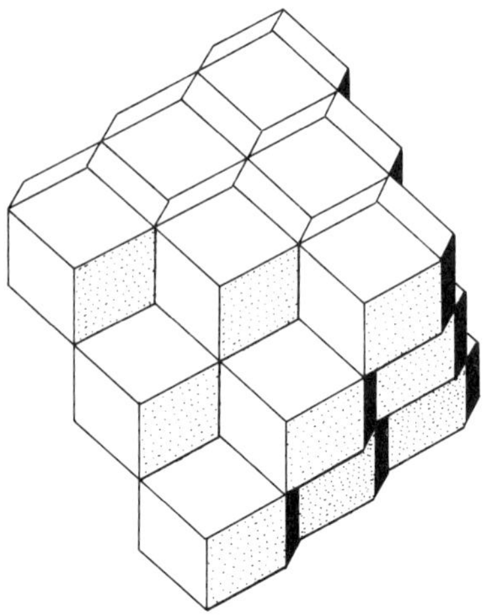

Fig. 172

The two interlocking R-D block puzzles above (Figs 171 and 172) are both surprisingly difficult to solve. Even if the reader discovers the design by experimental dissection or some other method and makes a set of pieces, the solution has a way of vanishing from memory the moment the pieces are scrambled. Those made with spheres might be even more confusing.

Chapter 19

Intermezzo

Perhaps this would be a good point at which to pause for a moment and review our progress thus far. Practically every puzzle design described in this book might be regarded as a systematic dissection of some geometrical form into bits, usually identical, which are then partially recombined into puzzle pieces. The superficial perception of this strange pastime is that all of this is so that a second party can then enjoy the confusion of trying to reconstruct the original solid. In point of fact, there is often no clear dividing line as to where the design process stops and the solution begins, or who is the designer and who is the solver. They may be one and the same. In some of the plane dissection puzzles in chapter 1, discovering the dissection is the real puzzler, after which the pattern solutions are relatively easy. In those like the Octahedral Cluster Puzzle, as presented, the design becomes the solution. In the Jupiter Puzzle, the intriguing design overshadows the straightforward solution because it is much the more interesting of the two. Some puzzles foisted upon the reader in previous chapters (if only one could write in a whisper) may not even *have* solutions!

It is common practice in most puzzle books to include the solutions somewhere. Perhaps some readers will be annoyed not to find them in this book. Solutions are fine when they serve some purpose. Certainly a book of riddles would be dull reading without the clever answers included, while answers to crossword puzzles may be educational. In the case of most combinatorial puzzles, including the solutions would add nothing new or interesting. There are exceptions, and note that some solutions have been included in the text. Here is a whole paragraph more of them:

Karl Essley's two misplaced pieces were of course identical and triangular, but we will never know who got them, will we? In the puzzling pairs of Tangram figures in which one appears to have a piece missing, the pieces are always rotated 45 degrees from one to the other. The mini-Tangram set of five pieces (Fig. 13) forms seven convex figures, five of which have multiple solutions. Beeler's proof of the impossibility of a 3×20 rectangular Cornucopia solution

(Fig. 39) is to count the empty squares on either side of the piece placed and note that they are never divisible by six.

In Fig. 41, the reader was asked to judge which of the two Cornucopia patterns was the more pleasing and to identify four flaws in the other one. The following is of course not an answer but merely collective opinion. See if the reader agrees. Of the half-dozen persons polled, all preferred the pattern on the right. One objectionable feature of the pattern on the left is the long horizontal line that nearly dissects the pattern. Another is the vertical line intersecting it and creating two 'crossroads'. (Could it be only a coincidence that a fundamental rule of good stone masonry construction is to avoid both long straight lines and crossroads for reasons of structural strength?) The two long parallel pieces at the bottom are also a distraction. A fourth flaw is that the first three flaws are all asymmetrical, creating a sense of unbalance. Having determined that this pattern is bad, it is interesting how many other objectionable features reveal themselves. The four vertical lines at the top lead the eye off the square, the *T* is upside down, the piece at the upper left is a pointing gun, and so on. Do you sometimes wonder what goes on (and off) inside the human mind?

What some readers may find even more perplexing than omission of solutions is that in many cases even the designs themselves are not shown in this book but instead left for the reader to ponder. The reason of course is that the design *is* the puzzle, so why spoil it by giving the answer. Publishing everything known on a subject may be a good idea in some fields, such as medicine. But in recreational mathematics, a gluttony of information is probably worse than none at all. With only a few exceptions, the policy in this book has been not to include the details for any puzzle designs or solutions that have not previously been published. Instead, they have been left purposely in the dark so that the inquisitive reader may have the joy of rediscovering some of them. This book is intended to be merely a glimpse into the puzzling world of polyhedral dissections and not an open pit excavation. If every known geometrical recreation were to be dug up, extracted, and refined, would it not leave a rather barren landscape behind for future generations?

The very idea that mankind might possibly be better off with some knowledge left unpublished probably sounds as far-fetched today as did the notion a century ago that some parklands ought to be left undisturbed. The compulsion not only to publish but to be the first to do so pervades the academic world. Added to that, we now look forward with some apprehension to the day when all of this and much more will be stored and analysed to death in some gigantic computerized retrieval system, with all the answers instantly accessible at the touch of a keyboard. But answers to *what*?

Computers and puzzles

The use of computers is now becoming fashionable in the world of geometrical puzzles. For solving certain types of combinatorial puzzles, once the program is in place, computers can be millions of times faster than a human, and more reliable too. Several solutions mentioned in this book, such as those for the pentominoes, would probably not have been tabulated except by computer. Such exercises usually have no practical value other than simply as a programming challenge or to satisfy someone's curiosity. There is probably not a single puzzle in this book that could not be solved by computer if someone wanted to go to the trouble of writing a suitable program. Some lend themselves much more easily than others, and some would present horrendous difficulties.

Now the computer is also being used as a designer's tool. It was mentioned how the computer saves time in checking out new design ideas for the six-piece burr, and how Cutler's computer-aided tabulation of burrs led to the illumination of two interesting versions that had lain dormant. The Cornucopia project was from the start an exploitation of state-of-the-art computer technology to compile a library of unique puzzle designs, which would have been impractical even just a few years earlier. A computer can even be instructed to search for most pleasing designs on the basis of certain aesthetic criteria, such as long lines and crossroads in Cornucopia solutions or difficulty index in burrs. But is this really aesthetics or pseudo-aesthetics? Is there any clear dividing line between the two, and are there any aesthetic qualities that a (non-human) computer, by definition, cannot be programmed to recognize and search for? Who knows even what is *really* meant by the word *aesthetics*?

The only significant advantage that a computer has over the human brain plus paper and pencil is blinding speed. Hence there is a tendency to program computers to solve combinatorial puzzles by brute force trial-and-error methods, whereas the human solver is always looking for clever shortcuts and usually finding them. This in itself can be a fascinating recreation. Solving geometrical puzzles by computer is rather like weeding your flower garden with a bulldozer. It may do the job quite thoroughly and rapidly, but consider for a moment all that is lost in the process, and what is the hurry in the first place?

In summary, computers are useful for solving problems that involve too much computation to be solvable by any other practical means or are just plain boring. They are misused for solving puzzles that we are either too lazy or too stupid to solve otherwise.

Abstraction and reality

Shown in Fig. 173a is a portion of a checkerboard dissection with x, y coordinates added. Any single square may now be designated by its x, y coordinates, and any puzzle piece by a group of such squares. Thus, the shaded piece is 1,1; 2,1; 2,2.

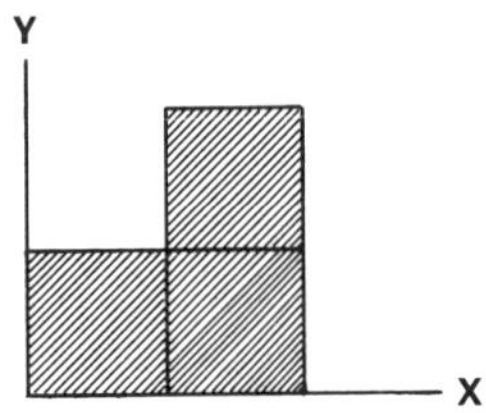

Fig. 173a

Given this notation (or some other of your liking), pieces may be moved about, rotated, turned over, fitted together, etc., all with numbers alone and with no need for the physical pieces or even drawings of them. This dimensionless world of numbers is of course the only world known to electronic computers. All puzzle problems must be reduced to it before being fed in, and any geometrical figures desired must be reconstructed after digestion and disgorgement by the computer.

It is easy to add a third dimension to this scheme and thereby use it to describe polycube puzzles. The puzzle piece shown in Fig. 173b would then be described in x, y, z coordinates as 1,1,1; 2,1,1; 2,2,1.

Such pieces may likewise be moved about and assembled analytically. Now the question arises, given the geometrical model and its numerical representation, which is the real puzzle and which is the abstraction? To pursue that question, consider the case of higher dimensions. This numerical notation works equally well in any dimension. A three-block piece in four dimensions—w, x, y,

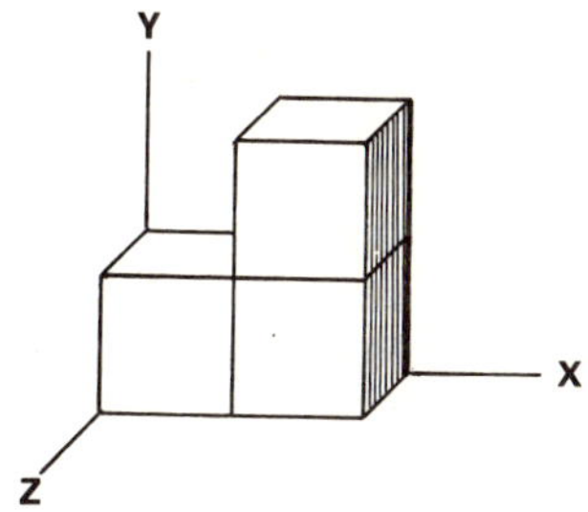

Fig. 173b

z—might be represented by 1,1,1,1; 1,1,1,2; 1,1,2,2. Note that each square in two dimensions is adjacent to four others, represented by adding or subtracting *one* from any one coordinate. Likewise, a cube in three dimensions is adjacent to six others, a block in four dimensions to eight others, and so on. Such higher dimension pieces may likewise be moved about and assembled into solid symmetrical solutions. The intriguing question of determining what would be considered 'interlocking' or 'assemblable' in four or more dimensions is left to the reader.

Now, which is the reality—numbers that we can understand (perhaps) and easily manipulate or hopelessly unimaginable hyper-geometrical models? Some Greek mathematicians, Pythagoras especially, were said to have regarded pure numbers alone as the ultimate reality in the universe and everthing else as a state of mind. Modern knowledge in neurophysiology and computer science casts this profound idea in a new light. Recent developments in theoretical physics go even farther into the abstract world of numbers, where physical models actually become utterly meaningless. Perhaps more to the point, what do the terms *physical* and *abstract* really mean, if anything?

The universal language of geometrical recreations

There must be very few if any other artefacts in the arts and sciences having the capability of transcending cultural barriers as do geometrical recreations. Show a dissection puzzle to persons anywhere in the world (or beyond!) and they are likely to grasp its simple message and start playing with it. Consider also their time-lessness. Anyone who spends much time pondering the mysteries of the polyhedra must sense a profound kinship with past cultures that have likewise come under their spell. Are we not all Pythagoreans?

Did children of yet far more ancient times fit together clay blocks into toy pyramids or, more likely, walls and fortifications of geometrical design? Gazing in awe into the star-studded sky, one can only wonder if other cultures in other worlds ponder these same geometrical puzzles.

The educational potential of geometrical puzzles does not seem to have been very fully exploited. A fascinating course in maths and logic could be constructed around some of the puzzles in this book. At the same time, think of all the other related subjects that could be tied in with it—history, art and sculpture, manual arts, philosophy, psychology—perhaps even the rudiments of Freudian analysis!

Games

Games and puzzles are closely associated. Sometimes the two words are used interchangeably, and the patents tend to be mixed together too. The most popular games have been board games, now being rivalled by video games, both of which are essentially two-dimensional. Devising a successful game that is truly three-dimensional has proven to be an elusive goal for many an inventor. There are certain practical difficulties in moving pieces about, adding or removing them in polyhedral space.

But the difficulties of polyhedral games go much deeper than that. Competitive amusements, by their very nature, tend to systematically exclude all irrelevant aspects of the game, *especially* aesthetics. Trying to devise a captivating game that also has much appeal to one's artistic sensibilities is almost a contradiction. Find one example if you can.

Many popular competitive games of today involve the symbolic capture, dominance, elimination, or destruction of one's opponents and their belongings. A favourite theme of video games is to accomplish this by blasting them to smithereens using the latest and most advanced space-age military weaponry. What the ultimate psychological consequences of all this may prove to be, no one really knows, but it is difficult to imaging doing that artistically. Next Christmas, why not instead give a child a hand-crafted burr puzzle or set of polyominoes. After all, someone had better begin practising how to put all the pieces back together again.

The whole idea of adults inventing games for children needs to be questioned. I used to try to devise games for our children, but I soon found that, given a box of wood scraps or other similar treasures, they would quickly invent their own simple amusements which they had more fun with than any of mine.

Chapter 20

The Two Tiers Puzzle

Some of the material for this book came from the far corners of the world and beyond. The following incredible story was provided by a correspondent in Calcutta named Paul E. Gahn. I cannot vouch for its authenticity and present it here just as it came to me.

Paul is a puzzle maker and puzzle historian of sorts. He was in his workshop late one evening when there was a knock on the door and a very attractive young woman entered. She held a small bundle wrapped in a scarf and said she had something to show him. As she unwrapped the bundle, six wooden puzzle pieces tumbled out onto the workbench. Paul immediately became quite interested. She showed him that all six pieces were supposed to be made of eight blocks glued together, but one of them had a block missing. She asked him if he would repair the broken piece and then assemble the puzzle. He said he would be glad to if she would tell him more about the puzzle, such as its origin. She said she would, and this was her amazing story:

The puzzle first appears in the hands of a prosperous Afghan merchant in the early 19th century, and where he got it no one knows. The merchant had two daughters. The older daughter fell in love with a British soldier, and soon they announced they wished to be married. The father was not too keen on this, and he decided to put the young man's wits to a test. He gave the soldier the puzzle in pieces and said that as soon as it was assembled the wedding could be arranged. The young man worked feverishly on it, with the girl secretly helping, and finally after two weeks proudly presented it to the father assembled. Plans were made for a wedding the following week. Just when everything seemed fine, the younger daughter brought the puzzle to her father and remarked that she could hear something rattling inside. Taking it apart, they found one block broken off and loose inside, and careful inspection revealed that it had been done in order to create a second solution. The enraged father summoned the two lovers and confronted them with the evidence, which they vigorously denied having any knowledge of.

At this point in the story, Paul interrupted and asked if it were not possible that the younger sister may have been jealous and had

Chapter 21

Theme and variations

One of the charms of the simple two-dimensional dissection puzzles shown in chapter 1 is that they construct many different simple geometrical shapes with the same set of pieces. Some of the polyhedral block puzzles in chapters 3 and 18 construct multiple shapes but they are non-interlocking. The difficulty of achieving this feature with interlocking puzzles was demonstrated in chapter 11 by a pair of designs that succeeded only partially. Is it possible for a set of interlocking puzzle pieces to construct many different polyhedral shapes?

The Six-Part Invention

Recall the simple two-piece dissection of the rhombic dodecahedron shown in Fig. 147. These half-pieces can be joined in pairs in different ways to form puzzle pieces. Excluding those that are impossible to assemble or have an axis of symmetry, there are 12 such pieces, shown in Fig. 175.

Let us make a list of possible constructions, i.e. ways that R-D blocks can be clustered symmetrically. To keep things simple, consider only those with six or fewer blocks. Eight such figures are shown in Fig. 176.

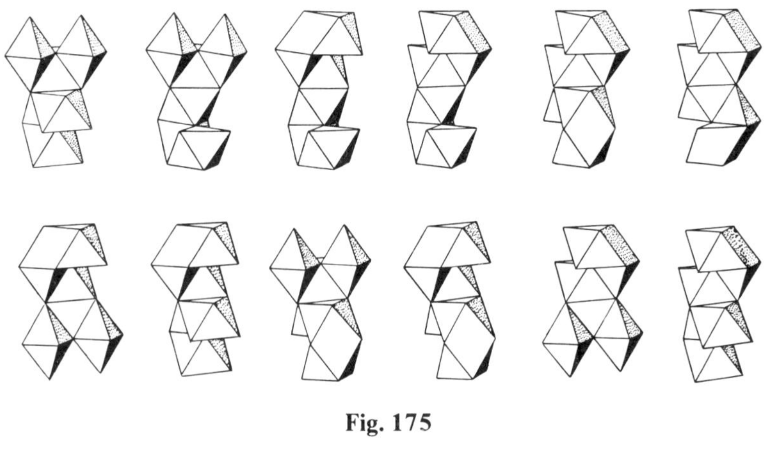

Fig. 175

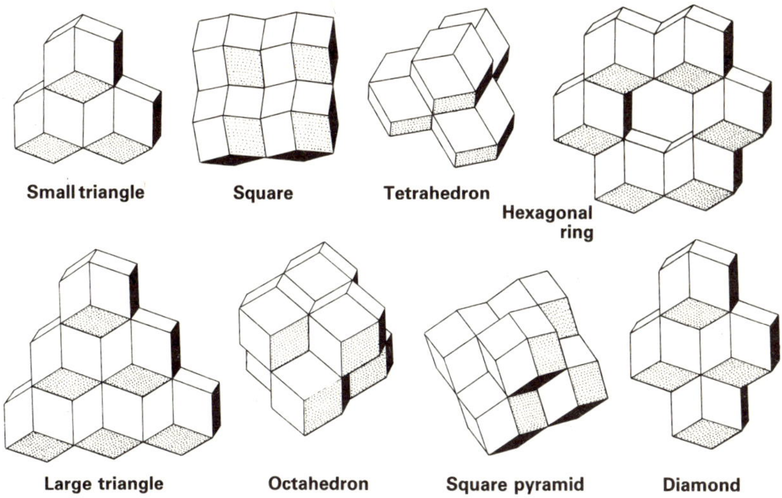

Fig. 176

Now for the interesting part. Can you find a subset of six pieces from the set of 12 that will construct all eight of the above figures? Do not be too discouraged if not, because Beeler's computer could not either. Of the 924 possible such subsets, there is one, however, that will construct seven of the eight figures. Find it if you can, keeping in mind that the seeking may be more fun than the actual finding.

Considering the thousands of different possible subsets of puzzle pieces and the many interesting constructions possible with any of them, the recreational potential for this family of puzzles is vast and practically unexplored. Choose your own personal subset of puzzle pieces and compile your own library of geometrical or animated shapes that they will construct (see Fig. 177). The pieces are also

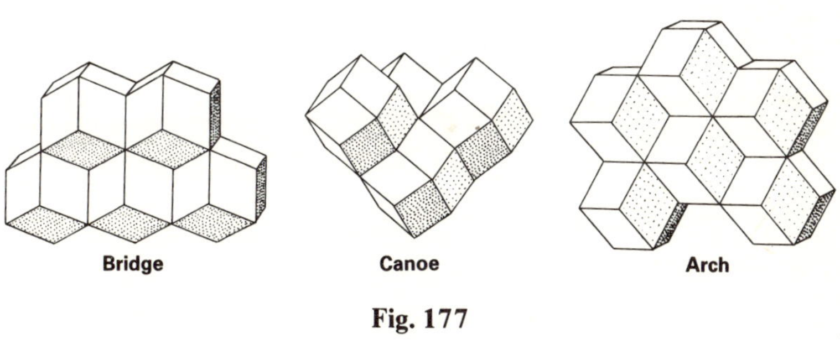

Fig. 177

great fun just to doodle with. Any closed loop can be considered a solution of sorts.

A well-crafted set of these pieces makes a most satisfactory puzzle. Each half of each puzzle piece is made of three squat octahedra building blocks joined accurately together, and the full pieces are then made up of these half-pieces joined different ways.

(Incidentally, to digress slightly, a fascinating recreation is to determine how many of the eight constructions shown in Fig. 176 are space-filling. You may be surprised at the answer.)

The Eight-piece Cube Puzzle

The design described above leads by analogy to one based on cubes in place of rhombic dodecahedra, dissected the same way. These half-pieces can be joined in pairs eight different ways, as illustrated in Fig. 178.

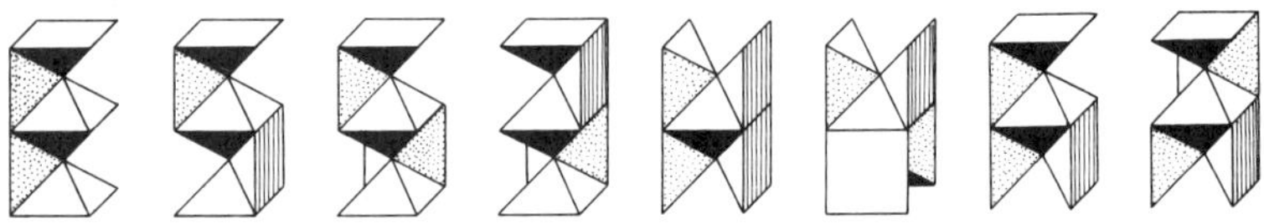

Fig. 178

The obvious question is whether these eight pieces can be assembled into a cube. That they do, and much more. Here is one of the best examples in this book of a geometrical recreation that lends itself to use in the classroom. For example:

1. Using two disconnected half-pieces, find all the ways that they can be joined face-to-face. You will of course arrive at the set of eight pieces, but this simple exercise can be quite instructive.

2. Prove that four pieces are the fewest that can be connected together in a closed loop. Prove that the square is the only such possible figure. Can two separate squares be made using all eight pieces? Why not?

3. Prove that the 2 × 4 rectangle is impossible. (Problems of this sort can always be solved systematically by trying every piece in every possible combination, but look for shorter and more elegant proofs using logic.) Now what other shapes cannot be made for the same reason?

4. Assuming all solutions to be closed loops, prove that an even

number of pieces must always be used. Find all possible solutions using six pieces. Likewise using all eight pieces. Examples are shown in Fig. 179.

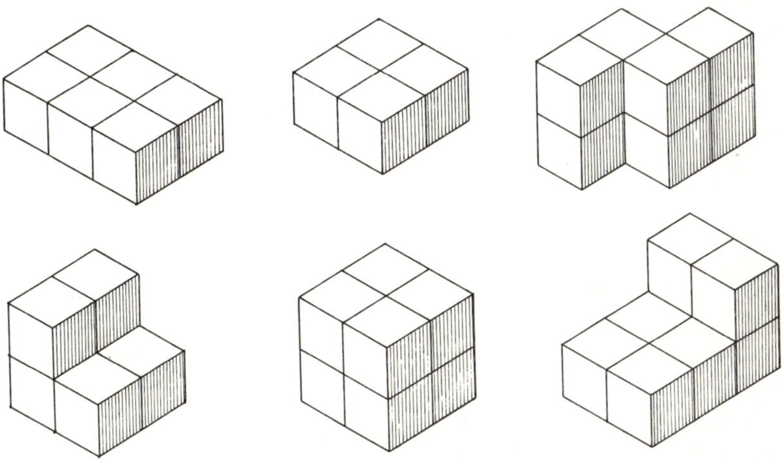

Fig. 179

Six of the pieces have reflexive symmetry and the other two are a reflexive pair. It necessarily follows that every solution must either be self-reflexive or occur in reflexive pairs. (These pairs are not counted as separate solutions.) Can you figure out why?

Some of the most fundamental questions in physics have to do with symmetry, and perhaps this puzzle will stimulate the student's interest in this fascinating subject. If the most elementary particles in the universe and all of the laws governing them were symmetrical (which is not to say they are), would it not follow that everything made from them, from atoms to the entire universe, should be either self-reflexive or one of a possible reflexive pair? But therein lies a curious paradox. Imagine that in the next instant the universe switched to its mirror image. How could you tell? Would not human consciousness be reflexive also? (Whatever *that* means!)

Another strange case is the DNA molecule and the genetic code. Most of us are right-handed, nearly all of us have our appendix on the right, and all of us carry DNA with a right-handed twist. How are instructions for right-handedness carried genetically? Would an identical but reflected DNA molecule produce an identical but reflected organism? A lucid discussion of these and many other fascinating problems in symmetry may be found in *The Ambidextrous Universe* by Martin Gardner, but don't expect to find all the answers.

The half-pieces for the Eight-Piece Cube Puzzle are made from three square pyramid blocks joined together. These blocks are made from sticks of isosceles–right-triangular cross-section with two 45-degree cuts. For experimental work, the mating joints can be slightly on the loose side. A more accurate model of this puzzle made of fine woods with close-fitting joints is a delight to play with. The sharp edges may be bevelled or rounded slightly to give its stark Bauhausian functionality a little more softness and warmth.

More variations

Although a well-crafted set of puzzle pieces for either of the two designs just described can be quite entertaining in itself, more important for the purpose of this book is that the geometrical principle they are both based on is even more fun to play with. It leads along an endless trail of new discoveries. For example, as suggested by Fig. 147, an obvious variation of the Six-Part Invention is to use the connection with three prongs rather than two. The three sample pieces shown in Fig. 180 assemble into a triangular cluster.

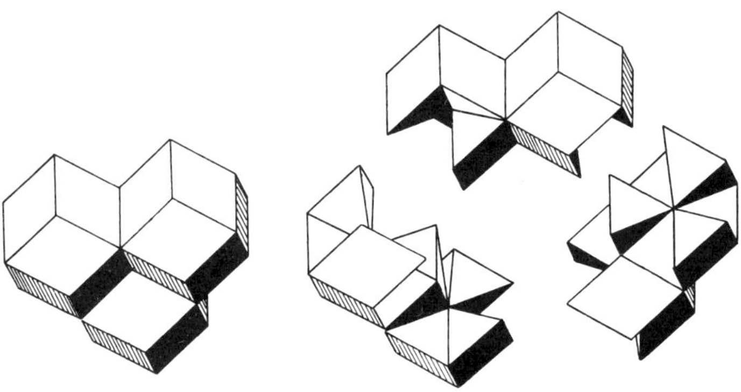

Fig. 180

By truncating the Eight-Piece Cube pieces to convert them into cuboctahedra, the puzzle remains functionally the same but assumes an intriguing new geometry (Fig. 181).

When any of the half-pieces described in this chapter are joined in threes rather than pairs, the numbers of puzzle pieces, practical

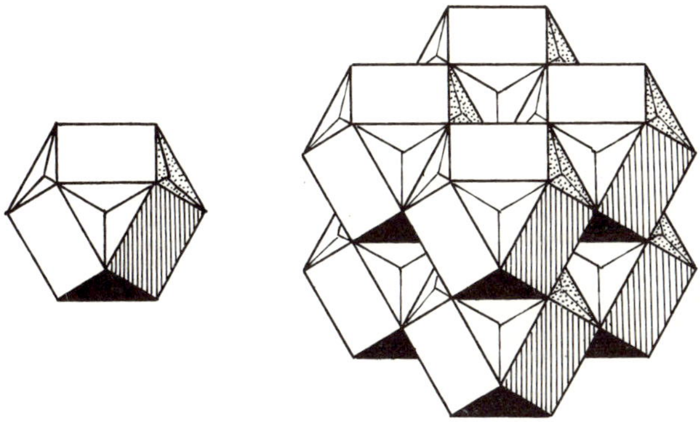

Fig. 181

sets, and possible constructions stretch the imagination. To give but one example, 12 identical pieces assemble to form the Triple Cross Puzzle shown in Fig. 182. Could you assemble 14 such pieces in axial symmetry? How about 46 such pieces?

Now imagine combining all of the above into one super set containing singles, doubles, and triple pieces, and perhaps some even larger. Simply as a play construction set, who could possibly resist the urge to tinker with these pieces and fit them together different ways? At the same time, this versatile set of puzzle pieces contains practically unlimited potential as an educational tool and as a kit for discovering new puzzle problems, some of which should baffle experts. A few sample puzzle constructions are shown in Fig. 183. A large set of such pieces, well-crafted of hardwood, makes a marvellous construction set.

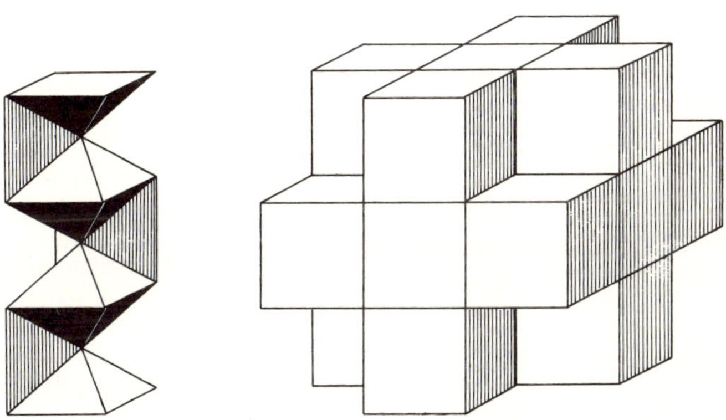

Fig. 182

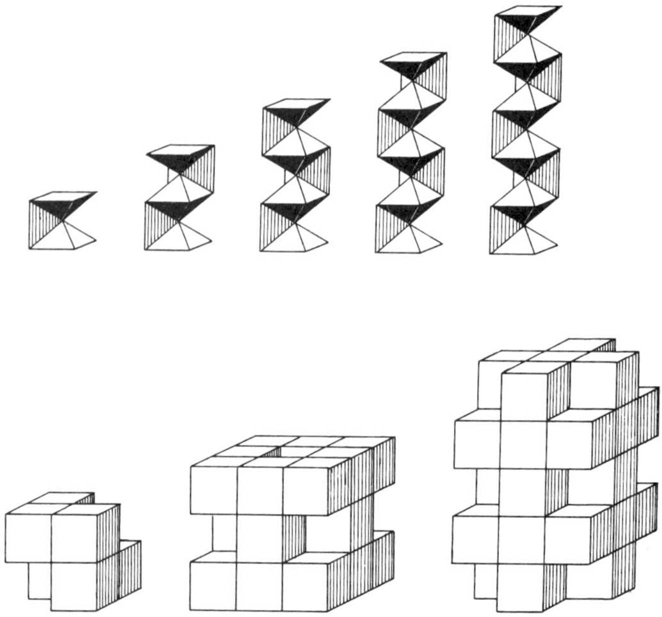

Fig. 183

The Pillars of Hercules family

When dissected cubes are combined with whole cubes, the number of puzzling possibilities takes another quantum jump. This idea hatched just in time for inclusion in this book, so just one simple example will be given here.

The set of seven dissimilar puzzle pieces shown in Fig. 184 could be described as three ordinary pentacube pieces and four jointed polycube half-pieces. They assemble one way only to form a $3 \times 3 \times 3$ cube. The next time someone says that $3 \times 3 \times 3$ cube puzzles are too easy, give them this one disassembled. Note that the half-pieces can be combined in pairs three different ways, with each pairing having four variations, so right there is ordinary puzzlement multiplied by a factor of 12!

Now, if you are the type that likes to have a little extra fun sometimes, design a $3 \times 3 \times 3$ cubic block puzzle like the above, except having only one jointed piece and with a tightly fitting joint. When carefully made, these joints are practically impossible to detect. With the jointed piece assembled backwards, it looks like just an

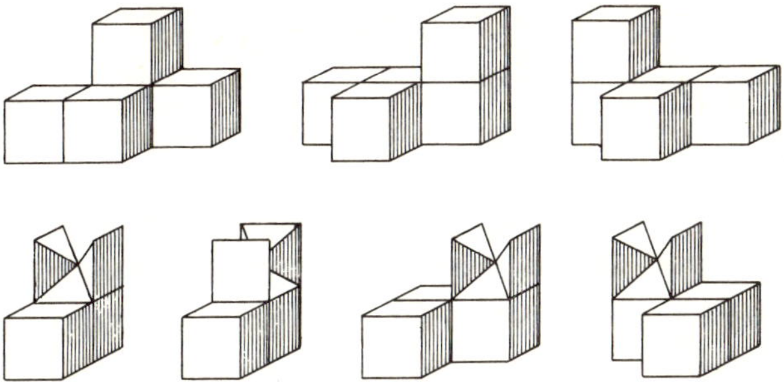

Fig. 184

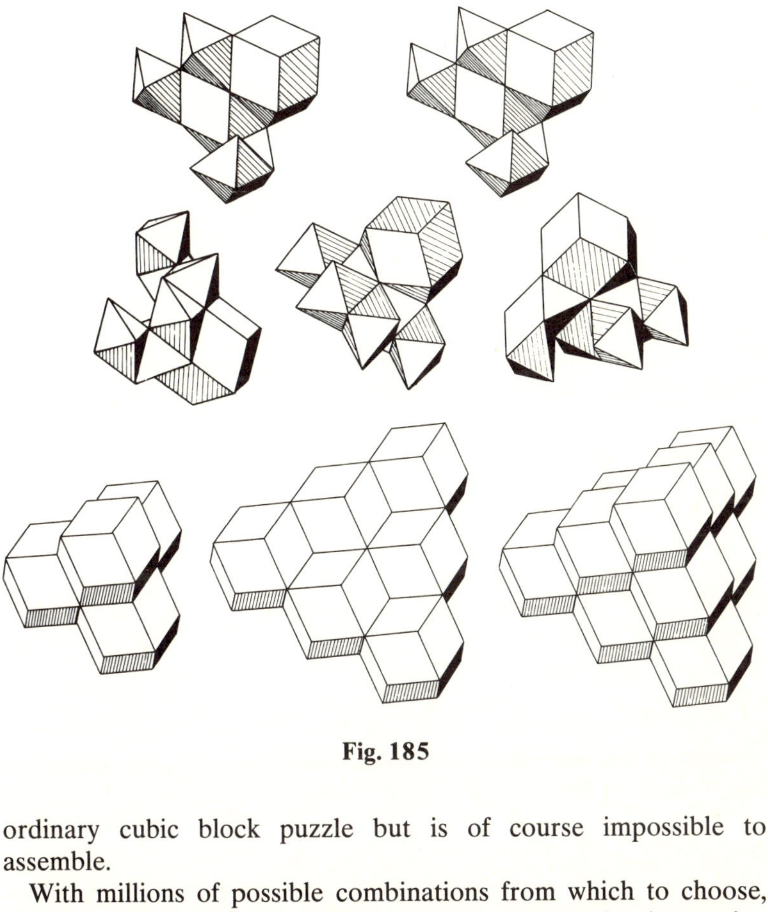

Fig. 185

ordinary cubic block puzzle but is of course impossible to assemble.

With millions of possible combinations from which to choose, clever designs of this sort might incorporate many other interesting features such as interlocking assembly and construction of multiple symmetrical problem shapes. This leads into a whole new world of puzzling possibilities, almost totally unexplored, hence the name *Pillars of Hercules.* Note also that such puzzles are fairly easy to make in wood, since most of the blocks are just cubes. More intriguing still is the possibility of extending this idea to rhombic dodecahedra, with whole blocks and half-blocks combined, as suggested by the sample pieces shown in Fig. 185. Two of them form a small tetrahedron, three of them a triangle, and all five a large tetrahedron.

Chapter 22

Blocks and pins

Most of us have at some time in our lives enjoyed playing with those marvellous construction sets consisting of blocks with holes joined together with dowels. Many interesting variations of these are possible. In three dimensions, the simplest and most obvious are cubic blocks with holes centred on their six faces, and with dowels all the same length or perhaps in integral multiples (see Fig. 186). These are easy for the home craftsman to make. Blocks of about one-inch size can be sawn out or purchased. Quarter-inch dowels may be found in most hardware stores. The ends of the dowels are slotted with a saw and the holes are drilled slightly undersized for a tight fit. Great accuracy is not required in drilling the holes, but a spur bit and a non-grainy wood such as maple will prevent the drill from wandering.

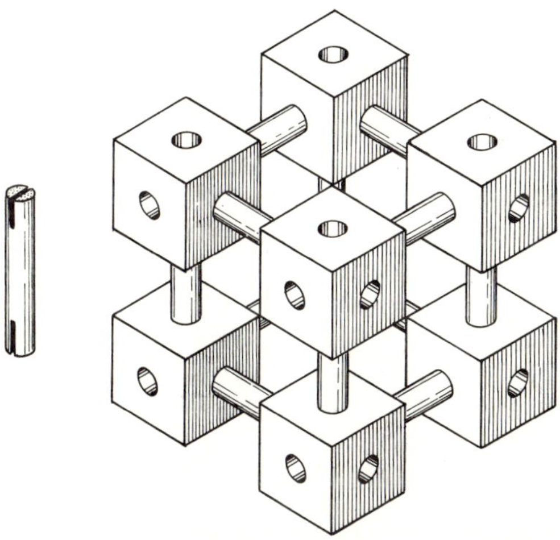

Fig. 186

A most interesting variation is to use edge-bevelled cubic blocks with 12 additional holes, as shown in Fig. 187. Many intriguing non-orthogonal geodesic constructions can be made with a set of these. The dowel lengths will be in multiples of $\sqrt{2}$.

Fig. 187

Another interesting variation is provided by the truncated cube (or truncated octahedron) with its eight additional holes. These will employ dowels having lengths in the ratio of $\sqrt{3}/2$, as in Fig. 188.

Finally, one might combine all of the above into a super-set of blocks, with all or some of the blocks having all 26 holes, together with dowels in all the appropriate lengths. One such block is shown in Fig. 189.

Simply constructing geometrical forms with such a set of blocks and dowels can be entertaining and educational, with or without illustrations as a guide. They also have potential for puzzle problems. The seven pieces shown in Fig. 190 comprise all the ways that one block and one dowel or two blocks and two dowels can be joined, using 12-hole blocks. Can they be assembled to fit snugly into the cubic box? What other symmetrical forms will they construct?

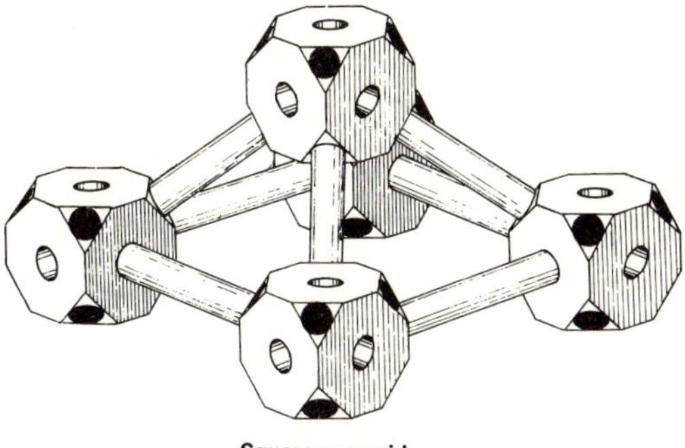

Square pyramid

Fig. 188

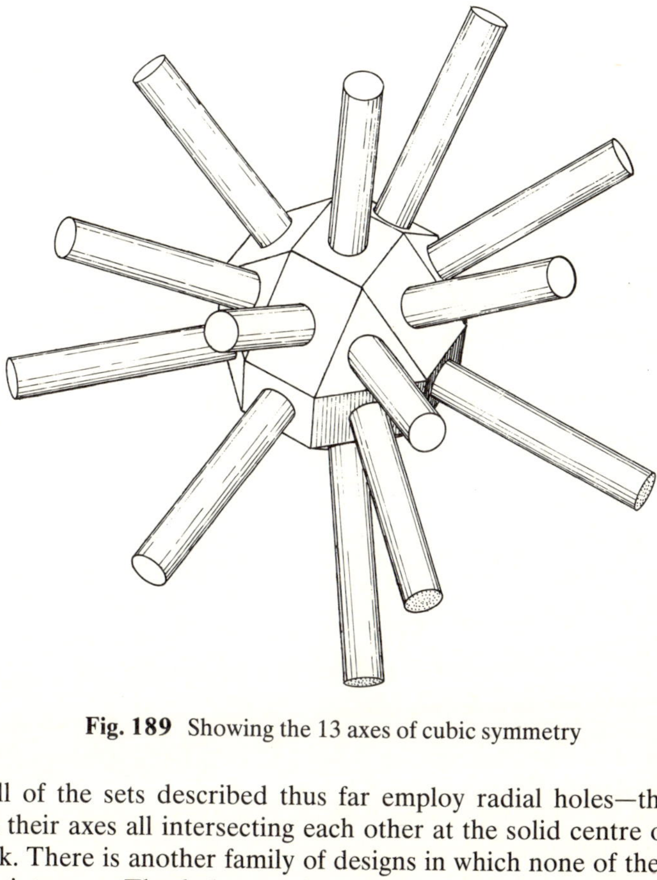

Fig. 189 Showing the 13 axes of cubic symmetry

All of the sets described thus far employ radial holes—that is, with their axes all intersecting each other at the solid centre of the block. There is another family of designs in which none of the hole axes intersect. The holes can be drilled straight through and the

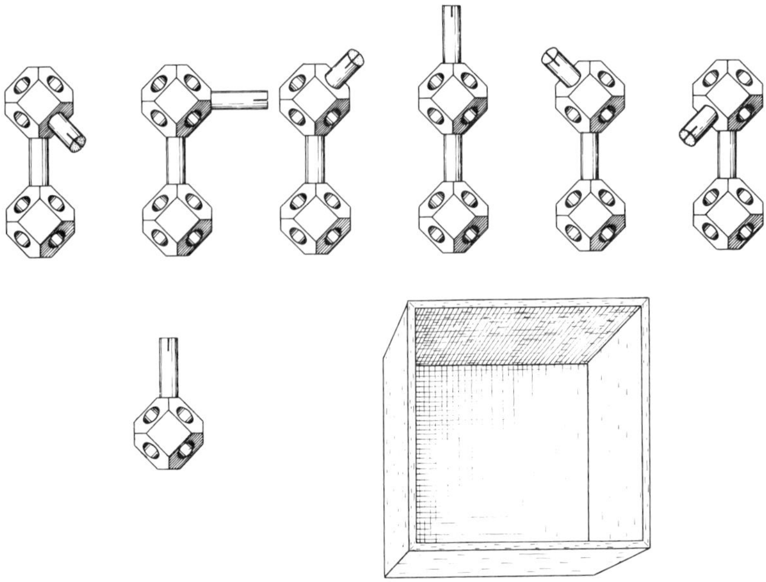

Fig. 190

dowels can be of indefinite length. The holes will be sized for the dowels to slide freely through.

This family can be further divided into two sub-families depending upon whether the drilled components are discrete blocks or uniform sticks of indefinite length. Examples of the latter have already been shown in chapters 6, 13, and 17. One further variation of the Pin-Hole Puzzle is shown in Fig. 191. The square sticks have holes equally spaced and arranged alternately. The

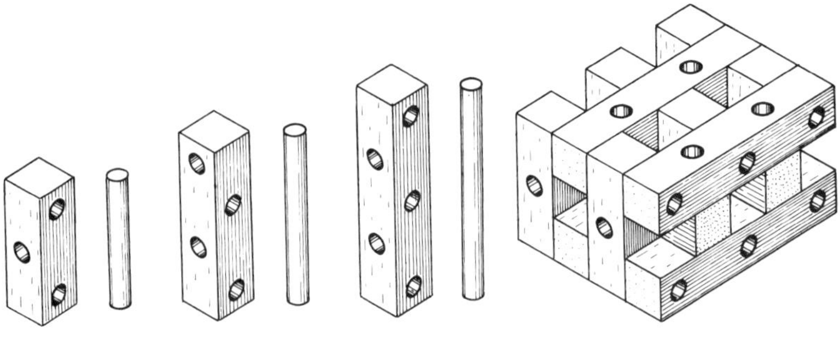

Fig. 191

assembly of sticks and dowels forms an orthogonal lattice that can extend indefinitely in all directions.

A set of such pieces might make a simple assembly plaything, perhaps to be fitted into a rectangular box. Or, with ingenuity, the idea might be developed into some sort of puzzle set. Incidentally, this entire chapter was conceived, developed, and rushed to the publisher just in time for inclusion, so many of these ideas are still 'in the rough'. (Are they not sometimes better that way?)

A cubic block with three mutually perpendicular non-intersecting holes drilled through it is shown in Fig. 192. It has a reflexive pair of forms.

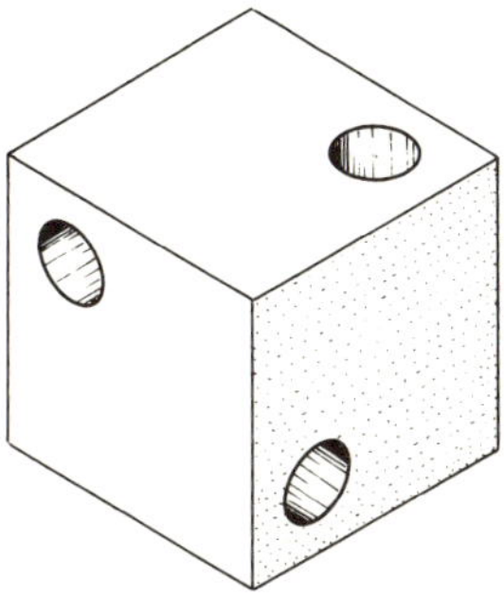

Fig. 192

An assembly of such blocks and dowels can be extended indefinitely. In the model shown in Fig. 193 on the left, all the blocks are identical. In the one on the right, the blocks alternate right-handed and left-handed. Note the two different types of symmetry that result. Only the assembly on the right can be said to have isometric symmetry as defined in chapter 6.

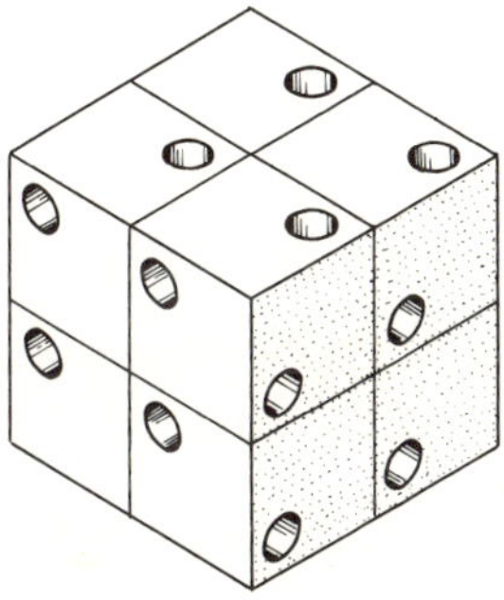

Fig. 193

Now consider the following problem: again start with a $2 \times 2 \times 2$ assembly of cubic blocks. Again drill three holes through each block so that 12 dowels can be inserted through the pile. But this time, all eight of the drilled blocks must be identical *and* the assembly must have isometric symmetry. The solution is shown in Fig. 194.

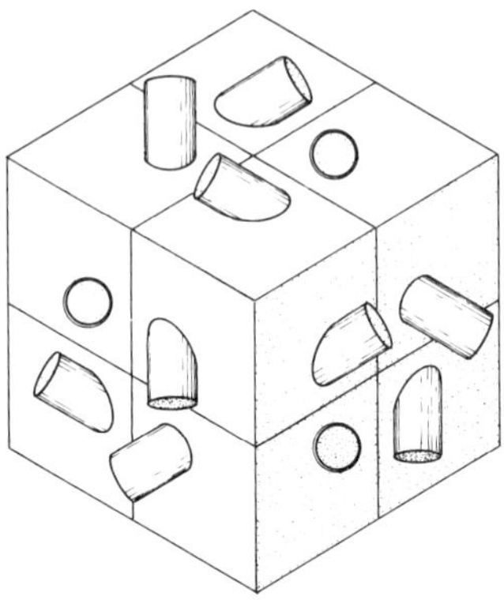

Fig. 194

The three identical holes in each block are all parallel to internal diagonals of the cube, and their axes exit the faces exactly one-third the distance from both edges. To be entirely satisfactory, the holes must be drilled accurately, and this will require a suitable jig set-up plus some patience to get it adjusted properly.

This is likewise an omnidirectional construction capable of infinite expansion in all directions. It has many fascinating variations. The $2 \times 2 \times 2$ grouping can in itself become a unit building block with 12 holes, or it could be broken into rectangular sub-units, as shown in Fig. 195.

The puzzling possibilities here would appear to be practically limitless. One interesting variation uses $1 \times 2 \times 2$ rectangular blocks. The four symmetrically arranged holes in each block pass diagonally through midpoints of sides. The blocks do not pack together but rather leave cubic and rectangular spaces. Neat symmetrical assemblies of six and twelve blocks are shown in Fig. 196.

A large set of such blocks and dowels is in itself fun to tinker with, but if some of the dowels and blocks are joined permanently and assembly problems are devised around them, they become intriguing puzzles as well.

In yet another variation, shown in Fig. 197 in two different views,

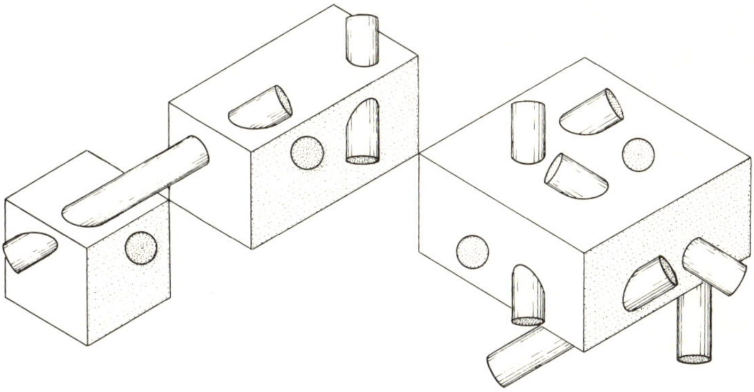

Fig. 195

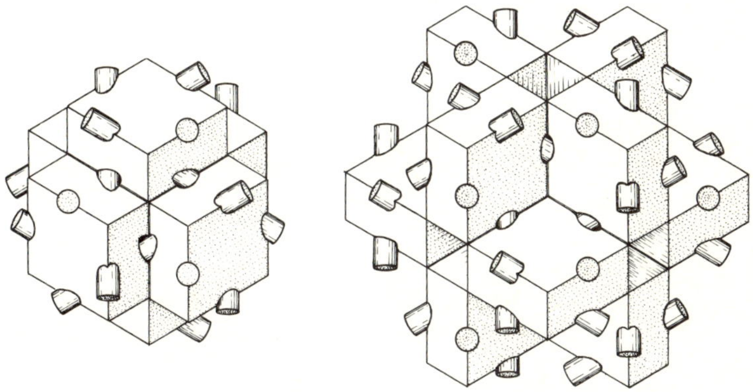

Fig. 196

squat octahedra have been substituted for the rectangular blocks above. Six of these are shown assembled with 12 dowels to form a solid rhombic dodecahedron. This construction is space-filling. Note that in the view along the threefold axis, the holes are centred in equilateral triangles.

Since the intriguing geometry of the rhombic dodecahedron is the basis for so many of the designs described in this book, combining it with our natural inclination for sticking pins into holes and joining things together should lead to many interesting new recreations.

By now, it should be clear that few, if any, of the designs described in this book could be considered novel inspirations except in some small part. They all are logical offspring of

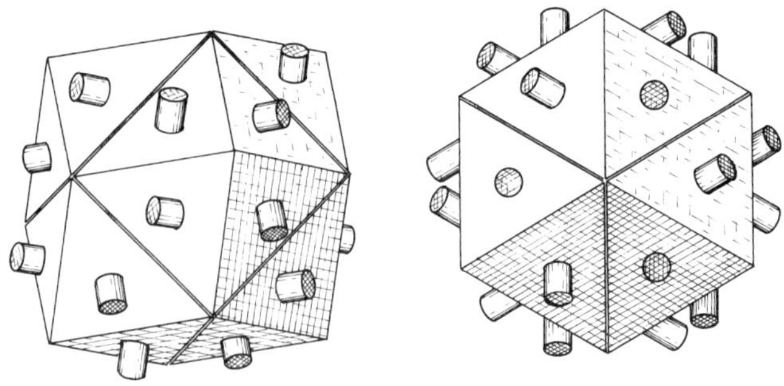

Fig. 197

previously established geometrical families, legitimate or otherwise. Mathematically speaking, the role played by the designer is often almost trivial. Once you start exploring this puzzling world of polyhedral dissections, one idea just leads to the next. Their arrangement in the book is an attempt to place the ideas in logical sequence of lineage. The problem is that one idea may have several roots branching backward in different directions. Often, one can arrive at the same place by two entirely different routes. Here is a good example:

Recall from chapter 13 the arrangement of 12 hexagonal sticks and dowels, shown on the left in Fig. 198. Now imagine that instead of the hexagonal sticks, all of the space surrounding the dowels is filled solid. Tessellate that space into space-filling rhombic dodecahedra, and dissect the central rhombic dodecahedron into six squat octahedra, shown in Fig. 198 on the right. The result is exactly the same as the design shown in Fig. 197.

The construction described above suggests compellingly by analogy dissection of other polyhedra into sections held together

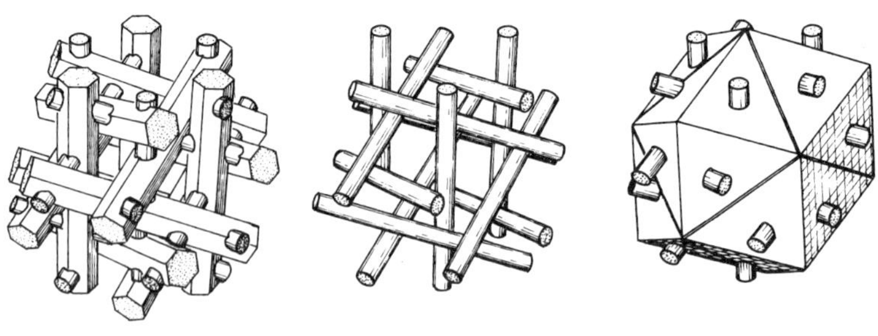

Fig. 198

with dowels. Shown in Fig. 199 is a stellated rhombic dodecahedron with 12 dowels drilled through it. There are many different practical ways that this solid might be dissected, such as into 48 tetrahedral blocks, 24 rhomboid pyramids, or 12 double rhomboid pyramids, as shown.

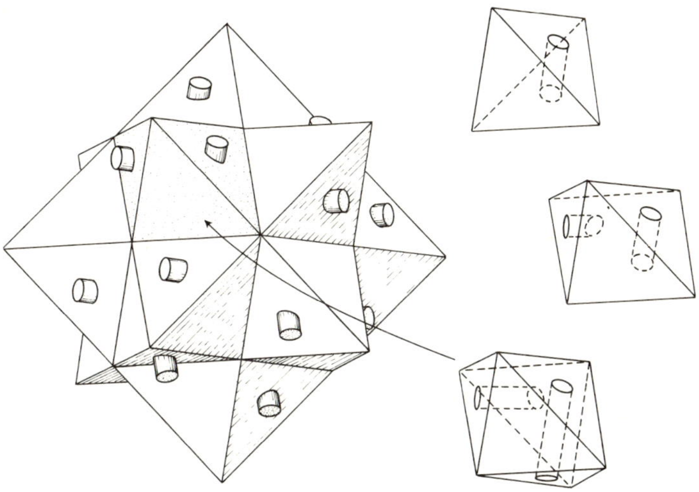

Fig. 199

Now compare the double rhomboid pyramid above with the squat octahedron of the previous design and note that they are the same geometrical solid with the same hole locations, the only difference being in the number of holes. Thus, a set of 12 four-hole blocks and dowels constructs two rhombic dodecahedra or one stellated version. If several of the dowels are fastened in place to form lollipop pieces (Fig. 200), assembly of these figures becomes an entertaining puzzle. What is the maximum number that may be joined and still be possible to assemble?

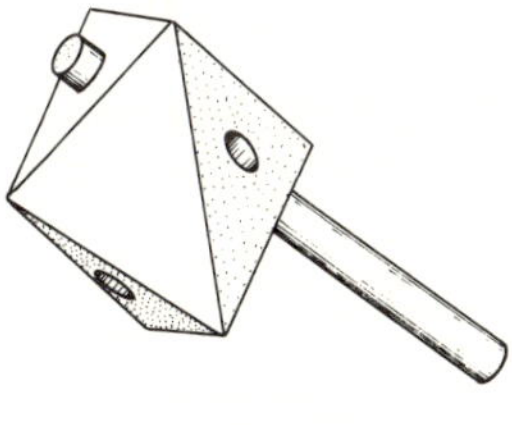

Fig. 200

These blocks are fairly easy to saw from square stock, as explained in chapters 8 and 9. A simple jig set-up can be used for the repetitive drilling of the holes. Since the holes are quite tilted toward the surface, a sharp spur point and soft wood are recommended to prevent wandering of the bit.

The Lollipop Puzzle

Carrying the scheme of the above design to its obvious next step, make the rhombic dodecahedron itself the basic construction unit. Each block will have 12 holes. Four such blocks are shown in Fig. 201, assembled into a tetrahedral pile and pinned together by twelve dowels.

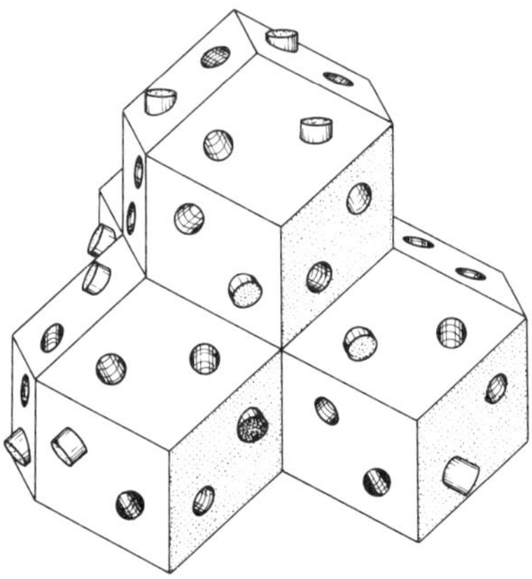

Fig. 201

In order to convert this intriguing construction set into an even more entertaining puzzle, we again join blocks and dowels to form lollipop pieces. But this time we also eliminate all extraneous holes. Not only does this save considerable drilling and improve the appearance, but it also adds greatly to the puzzling potential. By a judicious choice of hole locations, and stick locations in the lollipop pieces, simple constructions become fascinating puzzle problems. As a simple example, a triangular assembly of three blocks and three dowels is shown in Fig. 202. Each block has two holes. Two pairs of blocks and dowels are joined to form lollipop pieces. The remaining dowel is the key.

Note the similarities to the Pin-Hole Puzzle. The added feature of this new design is that even six or fewer rhombic dodecahedral

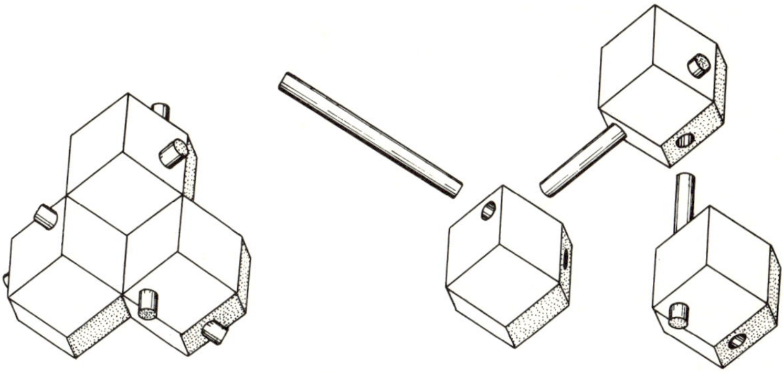

Fig. 202

blocks may be joined many different ways to create interesting geometrical shapes, as was shown for the Six-Part Invention in the preceding chapter. Thus, the natural appeal of pin-and-hole assembly pastimes is combined with the added feature of multiple assembly problems possible with one appropriately chosen set of puzzle pieces.

The arrangement of six blocks shown in Fig. 203 we have been calling an octahedral cluster. It uses 24 dowels. How many dowels may be attached to the blocks for it to still be assemblable? Can all the block pieces be dissimilar? What other puzzle problems can be

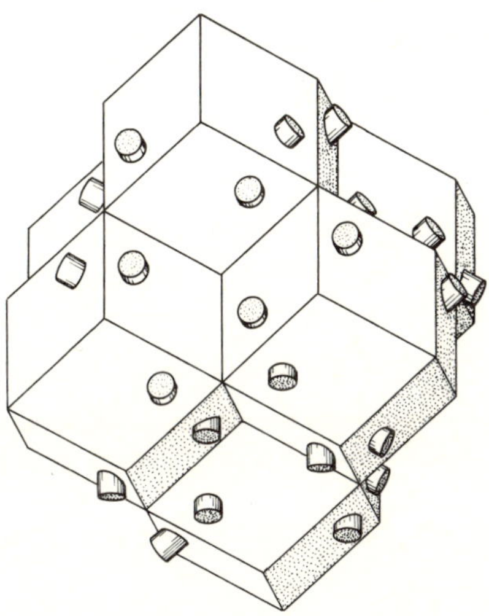

Fig. 203

devised using the same set of pieces? This should keep puzzle analysts busy for quite a while.

There is a limit to the diameter of the dowels in relation to the blocks. The diagram in Fig. 204 shows the limit to be 1:6. In other words, if the blocks are $1\frac{1}{2}$ inches across, the dowels cannot be more than $\frac{1}{4}$-inch diameter without interfering with each other.

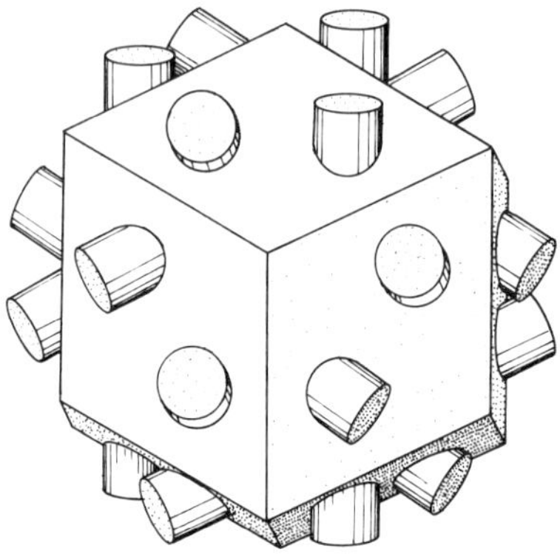

Fig. 204

If the dowels are made slightly larger in diameter than the limit shown above, a most interesting puzzle results. Some of the dowels will require cylindrical notches milled into them. Twelve such dowels are shown in Fig. 205 assembled inside a rhombic dodecahedral block with 12 holes. Note the similarity to the 12

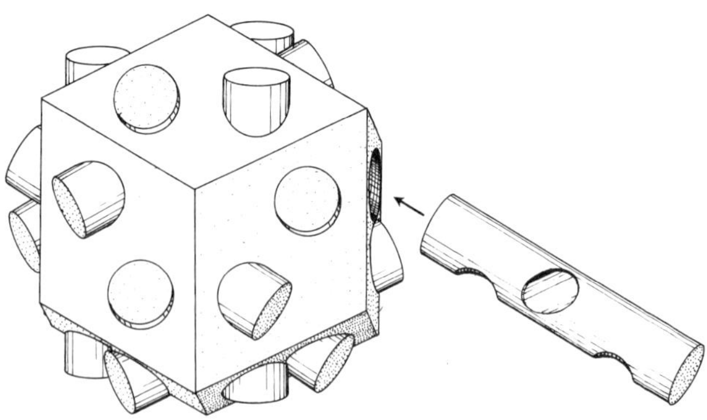

Fig. 205

notched hexagonal sticks on page 117, but with some added mechanical constraints. Also, you cannot see what is going on inside. Here is a case where clear plastic might be used to advantage for the block, and perhaps even for the rods too. This puzzle scheme likewise offers recreation for the designer as well as the solver, since many different notch combinations are possible.

Drawing on the brain

Incidentally, the reader may have noticed the changing style of illustrations used throughout this book, alternating between photographs and pen-and-ink drawings of varying quality. The explanation is that the author was in the process of experimenting with and learning both techniques, and many of the illustrations were redone again and again right up to the publication deadline in an attempt to improve both their clarity and visual appeal. Some of the diagrams in the first two chapters were redrawn by Jack Gray using a computer and laser printer.

Most of the polyhedral illustrations started out as photographs of models, using an ancient 4×5 Speed Graphic camera and retouching the enlargements with pencil or pen.

Anyone who has ever tried to compile a book of this sort knows that it is, more than anything else, a learning experience for the author. The discipline of organizing the information at hand invariably leads to new ideas that must be roughed out in the workshop, analysed, sorted and selected, with a few finished designs finding their way into the book as additions. This process of revision can become unending. Making accurate models, photographing them, developing, enlarging, and retouching is a time-consuming operation, and it tends to skew the text toward those that are easiest to fabricate, especially as publishing deadlines loom ahead. The alternative is to learn to draw one's inventions.

The graphic representation of geometrical models becomes fairly easy with practice and can be quite an interesting recreation in itself. The truth is, many of the designs shown here as pen and ink drawings, especially in the last three chapters, have never even existed as physical models as of the date of publication of this book, but rather only as abstract ideas transferred from the author's convoluted mind directly to the drawing pad. How remarkable that the human brain, given a two-dimensional graphic abstraction and a crude one at that, can in an instant mentally recreate to perfection the three-dimensional model that existed in the author's imagination—a process that might require in its physical form several days of painstaking work in a model shop by a skilled machinist or woodworker. Even more puzzling is that this bizarre geometrical re-creation should evoke so much pleasure for both the creator and the re-creator!

Chapter 23

Woodworking techniques

The techniques described in this book, and in more detail in this chapter, should allow the woodworker to at least make satisfactory experimental models, and with practice perhaps showpieces. Not much is required in the way of tools beyond a small power table saw and a drill press, without which, though, you will not be able to do a very good job.

Lumber

Satisfactory puzzles can be made from a variety of woods available at the local lumberyard. The lumber should be well seasoned, planed, and especially not badly warped. Poplar, basswood, fir, and pine are suitable woods for experimental work. Oak or cherry will produce a more finished product. For light and dark woods, as in checkerboards, birch and walnut can be used. For multicolour projects, fancy woods in a great variety of colours are available from specialty cabinet wood suppliers listed in any woodworker's magazine.

Most domestic hardwoods are quite unstable with changes in humidity, and this can be a serious problem with interlocking puzzles. Honduras mahogany and cedar are better in this respect. Some of the so-called exotic tropical woods such as rosewood, tulipwood, and ebony are better still; and teak, cocobolo, and padauk are best.

Forming into sticks

Every puzzle described in this book beyond chapter 1 is made from straight sticks, and well over half use straight *square* sticks. Making uniform accurate square sticks is by no means an easy task, especially when the lumber is slightly warped, as all lumber is. But it is absolutely essential to success. If the lumber is quite true and the saw is adjusted perfectly, you may be able to rip-saw the sticks straight away. Usually not all of these conditions are met, so you saw the sticks slightly oversized and then plane them down to exact size. Access to a small thickness planer will make this operation much easier. You can do many sticks at one time and save them for future use. If so, you will want to standardize on one or two sizes. One-inch-square sticks would be a good first choice. For puzzles using large numbers of cubic blocks, such as the solid pentominoes, $\frac{3}{4}$-inch is large enough. For a few, such as the truncated octahedra, you may want to use $1\frac{1}{2}$-inch-square stock. You will also require a measuring instrument. Vernier calipers are the absolute minimum, dial calipers are better, and if you are serious about this project, by all means use a micrometer.

Cross-cutting

The second and final sawing operation is to cross-cut these sticks into short stick segments or blocks. For this you will need a small table saw equipped with some special jigs. Much trouble will be saved if the saw makes a smooth cut that requires practically no sanding. The very fine-toothed so-called plywood blades do this. At the first sign of dullness they should be sharpened or discarded. Increased noise, resistance, or inaccuracy of cut are all signs of dullness; and when burn marks start to show the blade is hopelessly dull.

Nearly all puzzles are made of large numbers of identical blocks or sticks. The only way to saw these accurately and efficiently is by using special jigs that you make. Shown in Fig. 206 is a simple and very useful jig for sawing short square sticks and blocks. Its body is

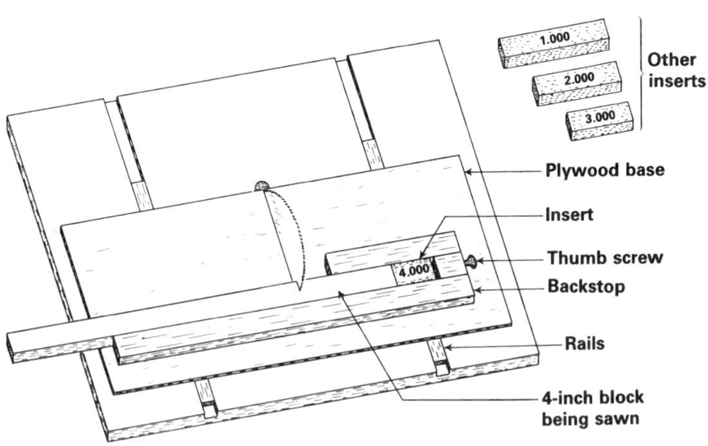

Fig. 206

a solid piece of plywood that slides on a pair of rails in the mitre grooves of the saw. Various inserts adjust the jig for making different sized blocks. The thumbscrew on the right allows for minute adjustments, such as might be necessary when changing saw blades. When correctly adjusted, all cuts should be accurate within plus or minus 0.005 inches.

A second very useful jig is the one already shown on page 78. With just these two jigs, one can make about half of the puzzles described in this book. A slightly modified version of the diagonal jig, shown in Fig. 207, is used to make rhombic dodecahedral blocks. As the stick is rotated to four different positions, four saw cuts are made, bringing the end to a pyramidal point. The stick is then advanced a certain distance determined by the spacer block, and four more shallow cuts are made. In the illustration, the final cut is being made that severs the finished block from the stock. Thus it is not necessary to place one's fingers near the saw. For one-inch stock, the length of the spacers is $\sqrt{2}$ plus the saw kerf.

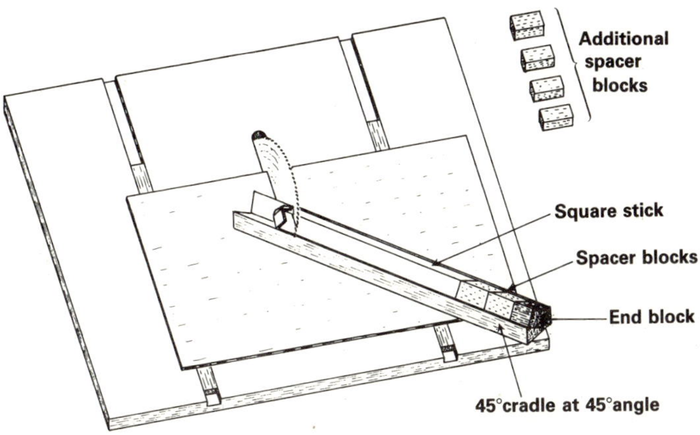

Fig. 207

A jig for sawing the eight corners from a cubic block to make truncated octahedra is shown in Fig. 208. The same sort of 45-degree cradle is used, but it forms an angle of $35\frac{1}{4}$ degrees to the mitre grooves when viewed from above.

Notched pieces such as are used in the standard six-piece burr are notched using the jig shown in Fig. 209 in conjunction with a dado blade in the saw. Spacer blocks are used to position the pieces properly.

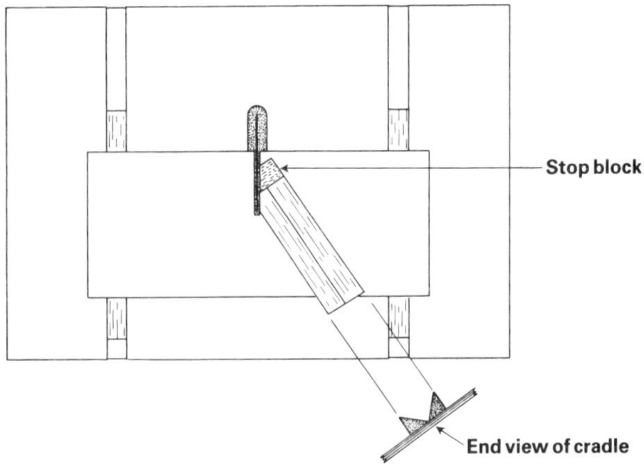

Fig. 208

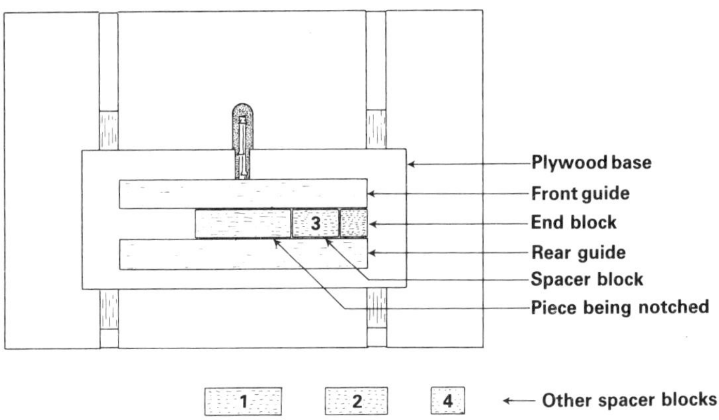

Fig. 209

For all of the puzzles that use equilateral-triangular sticks, the cuts are made using the simple jig shown in Fig. 210, which holds the sticks at an angle of $54\frac{3}{4}$ degrees viewed from above. Again, various spacer blocks are used to position the stock correctly.

There are many other similar special-purpose saw jigs, but the most basic ones have now been described and the reader should be able to figure out the others.

Drilling holes

For puzzles that use pins and holes, if the holes are drilled

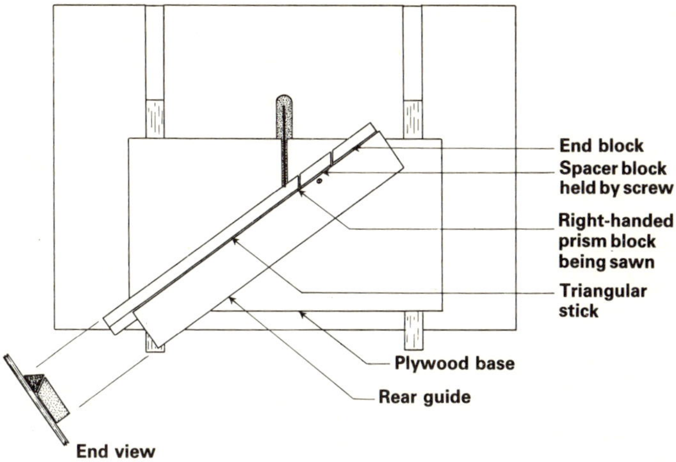

Fig. 210

accurately, you should be able to use a dowel $\frac{1}{32}$-inch smaller in diameter than the hole. Buy the dowel stock in any hardware shop. Always use a simple jig arrangement to position the pieces accurately while drilling—never just a pencil mark. It is best to use a special bit with a spur point made for wood rather than the ordinary bits that are designed for drilling metal and will not make a clean accurate hole in wood. Some woods are much easier than others to drill cleanly. Walnut is excellent.

Gluing

Most gluing is done using jigs to position the blocks accurately. A flat surface, straight edge, and combination square will suffice for gluing cubic or rectangular blocks. The simple M-shaped cradle shown in Fig. 211 is very useful and is used for making practically all of the puzzle pieces in chapters 8, 9, 10, and 11.

Many puzzles are most easily and accurately made by holding all of the blocks tightly together in the assembled configuration using tape or rubber bands and then selectively gluing them together. Examples would be the Three-Piece Block Puzzle, the Four-Piece Pyramid Puzzle, and the Octahedral Cluster Puzzle. Wax or bits of waxed paper are used to prevent unwanted joints from accidentally becoming stuck together. Sometimes, some of the joints can be sub-assembled first, with only the final joints being glued in the assembled shape, to ensure that the puzzle will be possible to

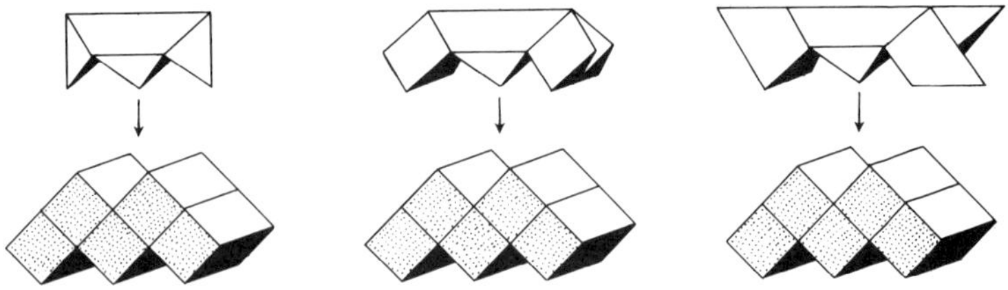

Fig. 211

assemble. For puzzles with multiple solutions, such as the Hexagonal Ring Puzzle, this method does not work and the only alternative is meticulous accuracy.

The most difficult puzzle pieces to glue are those of the Scorpius and Jupiter families. The base of the gluing jig for these is a vertex of a rhombic dodecahedron or triacontahedron. The author's were made by a skilled machinist using a milling machine. One pattern was used to cast a mould from which several more were cast in epoxy. The photograph (Fig. 212) shows one of the elves gluing Jupiter pieces.

The most satisfactory glue I have found is the yellow aliphatic resin type, sold under various brand names. It is strong, fairly fast setting, and resilient enough for the joints not to pop apart when the humidity changes.

Sanding and finishing

Interlocking puzzles such as the rhombic dodecahedral type of chapter 8 are most satisfactory when they fit snugly but not too tightly. The pieces are made and glued to be slightly too tight and are then carefully sanded down. The less sanding the better, as excessive sanding rapidly destroys the accuracy so carefully built in up to that stage. A belt sander with #150 grit is handy. Sometimes pieces can be exchanged between a puzzle that is too tight and another that is too loose. The last step is to break the sharp edges of the pieces with a file and round sharp corners using sandpaper. Thin clear lacquer applied with an artist's brush will improve the appearance and seal out dirt. For a final touch, rub with extra fine steel wool, wax, and buff.

Fig. 212

Summary

These brief woodworking hints are necessarily and purposely just that and not detailed directions. To give workshop blueprints for each puzzle would not only take a prohibitive amount of space but would also I think detract from the theme of the book. Part of the fun is figuring out how to do things. Because of the simple repetitive nature of geometrical dissections in general and the few recurring angles, the individual blocks are easier to saw out than one might suppose. By the same token, any error in sawing can become cumulative and surprisingly excessive when blocks are glued together. Accuracy at all stages of the work is most important.

Beyond that, the greatest problem in puzzle craft is coping with uneven expansion and contraction with changes in humidity. It can be as much as 2% from summer to winter. If it were uniform in all directions it would be no problem, but it can be 10 times as much across the grain as lengthwise. The result is that higher humidity tends to make most interlocking puzzles tighter. The burr puzzles are especially susceptible, and so the more stable types of woods should be used for them. Other designs, such as those of the Scorpius and Jupiter configurations, are barely affected by humidity—they just grow or shrink overall. By studying the geometry of a puzzle, sometimes one can find ways in the construction to minimize the effects of humidity. For example, in puzzles like the Three-Piece Block Puzzle or the Octahedral Cluster Puzzle, if the blocks are arranged such that their grains all run in the same direction in the assembled puzzle, the result will be an almost complete cancellation of the effects of humidity.

Start with the easier projects such as the two-dimensional dissection puzzles, the standard burr, and cubic blocks. As you gain experience with these and are able to make them to your satisfaction, then you may wish to progress toward the more difficult models. The arrangement of this book is roughly in order of increasing difficulty except where indicated otherwise. Do not be too disappointed if you find the Star Prism Puzzle or the Jupiter Puzzle to be beyond your woodworking capability. Curiously, it is the simpler puzzle that turns out almost invariably to have the greater recreational potential. For some indication of the author's personal preferences and recommendations, note the amount of space devoted in this book to each puzzle design.

For more information the reader is referred to my previous book, *Puzzle Craft*, which is oriented more toward the practical woodworking aspects of geometrical puzzles.

Finale

Truly, one of the most fascinating aspects of this puzzling world of polyhedral dissections is the strange counterpoint played by the duo of mathematics and aesthetics. Indeed, that is really what this book is all about. It seems that you can never consider one without the other. In fact, any reader who knows what the difference is between art and science, or can tell where one leaves off and the other begins, is way ahead of the author on that score. They both play the same haunting theme, and its echoes reverberate throughout Puzzledom like the proverbial horns of Elfland. To the Greeks, this elegant cosmic order was their fabled 'music of the spheres'. Perhaps this could be rephrased slightly to include the ubiquitous cube and the intriguing rhombic dodecahedron!

Whether one's main interest is in solving puzzles, collecting them, reading about them, or whatever, the reader is encouraged to try making at least a few of these models in the school or home workshop. In all likelihood, this will lead in turn to the ever more fascinating recreation of prospecting for new puzzle ideas. They lie scattered all about waiting to be uncovered, enjoyed, and shared with family and friends. It helps if you dig with the right tools. A few have been supplied in this book, but really the basic tools are imagination and curiosity. The essence of any creative endeavour is the fitting together of ideas into a harmonious whole, and the inspiration for that may come from almost anywhere.

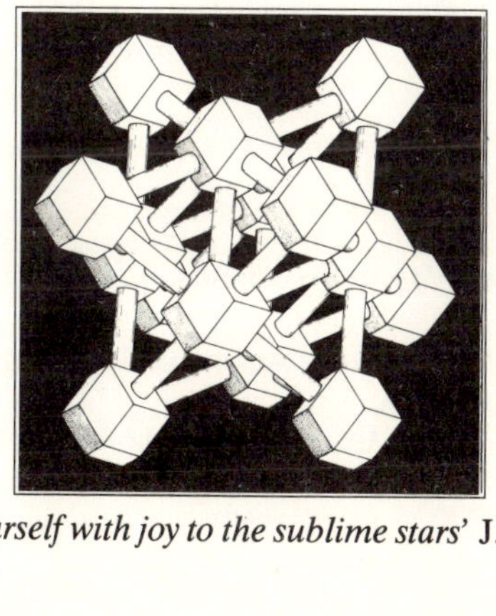

'Lift yourself with joy to the sublime stars' J. S. Bach

Bibliography

Books

Botermans, J. and Slocum, J. (1986) *Puzzles Old and New, How to Make and Solve Them*, Seattle, University of Washington Press.
Coffin, S. (1985) *Puzzle Craft*, Stewart T. Coffin, 79 Old Sudbury Road, Lincoln, MA 01773, USA.
Dudeney, H. E. (1907; 1919) *The Canterbury Puzzles*, London, Nelson. (Reprinted in 1958 by Dover Publications, New York.)
Elffers, J. (1979) *Tangram—The Ancient Chinese Puzzle*, New York, Harry N. Abrams.
Farhi, S. (1982) *Soma World*, Sivy Farhi, 815 S. California Unit B, Monrovia, CA 91016, USA.
Fourrey, E. (1907; 1938) *Curiosités Géométriques*, Paris, Vuibert.
Gardner, M. (1971) *Martin Gardner's Sixth Book of Mathematical Games from Scientific American*, San Francisco, W. H. Freeman.
Gardner, M. (1978) *The Ambidextrous Universe*, New York, Charles Scribner's Sons.
Golomb, S. (1965) *Polyominoes*, New York, Charles Scribner's Sons.
Hoffmann, Professor (1893) *Puzzles Old and New*, London, Frederick Warne & Co.
Kadon Enterprises, Inc., (1986) *Quintillions*, 1227 Lorene Drive, Pasadena, MD 21122, USA.
Lindgren, H. (1972) *Recreational Problems in Geometrical Dissections and How to Solve Them*, New York, Dover Publications.
Slocum, J. (1983) *Compendium of Checkerboard Puzzles*, Jerry Slocum, Beverly Hills, CA, USA.
Steinhaus, H. (1969) *Mathematical Snapshots*, Oxford University Press.
van Delft, P. and Botermans, J. (1978) *Creative Puzzles of the World*, New York, Harry N. Abrams.

Periodicals

American Mathematical Monthly, PO Box 6248, Providence, RI 02940, USA.
Journal of Combinatorial Theory, Academic Press Inc., 1250 Sixth Ave, San Diego, CA 92101, USA.
Journal of Recreational Mathematics, 120 Marine St., Farmingdale, NY 11735, USA.

Scientific American, New York, 415 Madison Ave, New York, NY 10017, USA.

World Game Review, 3367-I North Chatham Road, Ellicott City, MD 21043, USA.

Manufacturers

Kadon Enterprises, Inc., 1227 Lorene Drive, Pasadena, MD 21122, USA.

Pentangle, Over Wallop, Hants SO20 8JA, England.

Trench Enterprises, Three Cow Green, Bacton, Stowmarket IP14 4HJ, England.

Index